Handbook of Augmented and Virtual Reality

Augmented and Virtual Reality

Edited by
Vishal Jain

Volume 1

Handbook of Augmented and Virtual Reality

Edited by
Sumit Badotra, Sarvesh Tanwar, Ajay Rana, Nidhi Sindhwani
and Ramani Kannan

DE GRUYTER

Editors

Dr. Sumit Badotra
School of Computer Science
Engineering and Technology
Bennett University
201310 Greater Noida
Uttar Pradesh
India
summi.badotra@gmail.com

Dr. Sarvesh Tanwar
Amity University
Institute of Information Technology
Sector 125, Noida
201301 Uttar Pradesh
India
s.tanwar1521@gmail.com

Prof. Ajay Rana
Shobhit University
NH 58, Modipuram, Meerut
250110 Uttar Pradesh
India
askajayrana@gmail.com

Dr. Nidhi Sindhwani
Amity University
Institute of Information Technology
Sector 125, Noida
201301 Uttar Pradesh
India
manchandanidhi804@gmail.com

Dr. Ramani Kannan
University Teknologi PETRONAS
Center for Smart Grid Energy Research
Perak Darul Ridzuan
32610 Seri Iskandar
Malaysia
ramani.kannan@ieee.org

ISBN 978-3-11-078516-6
e-ISBN (PDF) 978-3-11-078523-4
e-ISBN (EPUB) 978-3-11-078531-9
ISSN 2752-2156

Library of Congress Control Number: 2023933966

Bibliographic information published by the Deutsche Nationalbibliothek
The Deutsche Nationalbibliothek lists this publication in the Deutsche Nationalbibliografie; detailed bibliographic data are available on the Internet at http://dnb.dnb.de.

Cover image: Thinkhubstudio/iStock/Getty Images Plus
Typesetting: VTeX UAB, Lithuania
Printing and binding: CPI books GmbH, Leck

MIX
Papier | Fördert
gute Waldnutzung
FSC® C083411
FSC
www.fsc.org

www.degruyter.com

Preface

AR/VR technologies, also known as Immersive Technologies, are revolutionizing the present and future technologies. These rapidly growing and fascinating technologies are used to create or extend reality and have applications in almost every field. The simulated environment using AR/VR for specific applications provides users with safe and affordable training. In case of emergency, these technologies offer expertise and preparedness for users. Moreover, AR/VR has tremendous benefits in education, with many major companies such as Microsoft, Apple and many others already making use of AR/VR. It will also revolutionize the man-machine interface in ways we could not have imagined 35 years ago when personal computers first emerged and changed our world forever. With AR/VR, there is no need to create an artificial environment; instead these technologies work with the already available environment or applications. They facilitate the design and development of 3D physical models, which are considered critical parts of the application design process. Software used for AR design, along with advanced visualization technologies of VR, can be very helpful in implementing and designing the ecosystem. In summary, we can say that AR is an additive technology that overlays created digital content onto the real world, providing an enhanced visualization experience. While making use of AR/VR technologies, users and can still interact with their existing environment. The visualization experience is affordable, which is one of the biggest advantages of AR/VR.

This book contains chapters dealing with various aspects related to applications and challenges in AR/VR, including:

1. AR/VR challenges
2. Recent trends in AR/VR
3. AR/VR applications in areas such as education, real estate, healthcare, manufacturing, media and more
4. Augmented Reality for robotics and Artificial Intelligence
5. AI methods and techniques for AR applications
6. Data visualization and analytics with Virtual Reality
7. Applications of IoT to enhance existing VR/AR/Mixed Reality devices/applications
8. Digital twins in industry and academia
9. Convergence of AR/VR with IoT
10. Enhancing AR with Machine Learning techniques
11. Use cases of blockchain for AR/VR

https://doi.org/10.1515/9783110785234-201

Contents

Editors' Profile

Dr. Sumit Badotra

Dr. Sumit Badotra is an Assistant Professor in the School of Computer Science and Engineering, School of Computer Science and Engineering, Bennett University, Greater Noida, Uttar Pradesh, India. He has around five years of teaching and research experience in Software Defined Networks (SDN). His general research interests lie in the areas of Network Security and Computer Networks, with specific research interests in Intrusion Detection and Protection from Internet Attacks. During his PhD work, he has served as a Research Fellow on a project funded by the DST Government of India. He has published over 60 papers SCI/Scopus/UGC approved journals, reputed national/international conferences and book chapters. He has filed several patents in relevant fields, attended numerous national-level FDPs and workshops, and acted as a resource person. Dr. Badotra is an active reviewer for various reputed journals. He is currently exploring Intrusion Detection in cloud-based web servers using SDN.

Dr. Sarvesh Tanwar

Dr. Sarvesh Tanwar is an Associate Professor at Amity Institute of Information Technology (AIIT), Amity University, Noida. She heads the AUN Blockchain & Data Security Research Lab. With over 15 years of teaching and research experience, her research areas include Public Key Infrastructure (PKI), Cryptography, Blockchain and Cyber Security. She has published more than 100 research papers in international journals and conferences. Dr. Tanwar is currently guiding six Ph. D. scholars and has supervised one Ph. D. scholar and five M.Tech research scholars. She has filed 21 patents (15 published) and two copyrights in the relevant field. She is senior member of IEEE, a Life-Member of the Cryptology Research Society of India (CRSI), the Indian Institute of Statistics, Kolkata, India and member of International Association of Computer Science and Information Technology (IACSIT) in Singapore. Dr. Tanwar serves as a reviewer in various reputed journals.

Prof. (Dr.) Ajay Rana

Prof. (Dr.) Ajay Rana is Professor of Computer Science and Engineering, currently serving as Director General in Amity University Uttar Pradesh, Greater Noida. With over two decades of experience in academics and industry, Dr. Rana completed his M. Tech. and Ph. D. in Computer Science and Engineering from reputed institutes of India. He has 117 patents under his name in the field of IoT, Networks, and Sensors. He has published more than 306 research papers in reputed journals and international and national conferences, co-authored bine books and co-edited 45 conference proceedings. Eighteen students have completed their Ph. D. under his supervision and currently six students are working on their doctorates with his guidance.

Dr. Nidhi Sindhwani

Dr. Nidhi Sindhwani is currently working at Amity Institute of Information Technology (AIIT), Amity University, Noida, India. She holds a Ph. D. (ECE) from Punjabi University, Patiala, Punjab, India, and has a teaching experience of over 15 years. Dr. Sidhwani is a Life-Member of Indian Society for Technical Education (ISTE) and a Member of IEEE. She has published three book chapters in reputed books, ten papers in Scopus/SCIE Indexed Journals and four patents. She has presented various research papers in national and international conferences and chaired a session at two international conferences. Her research areas include Wireless Communication, Image Processing, Optimization, Machine Learning, IoT etc.

https://doi.org/10.1515/9783110785234-202

Dr. Ramani Kannan

Dr. Ramani Kannan is currently a Senior Lecturer at the Center for Smart Grid Energy Research, Institute of Autonomous System, University Teknologi PETRONAS (UTP), Malaysia. Dr. Kannan completed his Ph. D. (Power Electronics and Drives) from Anna University, India in 2012, M.E. (Power Electronics and Drives) from Anna University, India in 2006 and B.E (Electronics and Communication) from Bharathiyar University, India in 2004. With over 15 years of experience in prestigious educational institutes, Dr. Kannan has published more than 130 papers in various reputed national and international journals and conferences. He is an editor, co-editor, guest editor and reviewer for various books, including Springer Nature, Elsevier etc. Dr. Kannan has received the award for best presenter at CENCON 2019, the IEEE Conference on Energy Conversion (CENCON 2019) in Indonesia.

List of Contributing Authors

Muhammad Makarfi Ahmad
Department of Agricultural Economics
and Extension
BUK
Kano
Nigeria

Arzoo
Department of CET
Guru Nanak Dev University
Amritsar
India
E-mail: arzoorana98@gmail.com

Maryam Babawachiko
MCA-Google Apps
Suresh Gyan Vihar University
Jaipur
India

Ajay Sudhir Bale
Department of ECE
New Horizon College of Engineering
Bengaluru
India
E-mail: ajaysudhirbale@gmail.com

Salil Bharany
Department of Computer Engineering & Technology
Amritsar
India
E-mail: salil.bharany@gmail.com

Vimal Bibhu
Amity University
Greater Noida
UP
India
E-mail: drvimal@gn.amity.edu

Sangeeta Borkakoty
Department of Computer Science & Electronics
University of Science & Technology
Meghalaya
793101
India
E-mail: s.borkakoty06@gmail.com

Kajal S Bundele
Department of CSE
School of Engineering and Technology,
CMR University
Bengaluru
India
E-mail: kajalbundele26@gmail.com

Khushi Dadhich
Amity University
Greater Noida
UP
India
E-mail: khushidadhich08@gmail.com

Lipsa Das
Amity University
Greater Noida
UP
India
E-mail: lipsaentc9@gmail.com

Medini Gupta
Amity Institute of Information Technology
Amity University Uttar Pradesh
Noida
India
E-mail: guptamedini642@gmail.com

Muhammed Furqaan Hashim
Department of CSE
School of Engineering and Technology,
CMR University
Bengaluru
India
E-mail: furqaanhashim@gmail.com

Daisy Kalita
Department of Computer Science & Electronics
University of Science & Technology
Meghalaya
793101
India

Gunseerat Kaur
Department of Computer Science and Engineering
Lovely Professional University
India
E-mail: Gunseerat.22157@lpu.co.in

Kiran Bir Kaur
Department of Computer Engineering & Technology
Amritsar
India
E-mail: kiran.dcse@gmail.com

Prithiraj Mahilary
Department of Computer Science & Electronics
University of Science & Technology
Meghalaya
793101
India

Rohan Mulay
SCIT
Manipal University Jaipur
Jaipur
India
E-mail: mr.rohan99@gmail.com

Radhika
Department of Computer Engineering & Technology
Amritsar
India
E-mail: kumariradhika.178@gmail.com

Ajay Rana
Amity University
Greater Noida
UP
India
E-mail: ajay_rana@amity.edu

Mohammed Sanusi Sadiq
Department of Agricultural Economics
and Extension
FUD
Dutse
Nigeria
E-mail: sadiqsanusi30@gmail.com

Bhuvi Sharma
Amity University
Greater Noida
UP
India
E-mail: bhuvisharma86@gmail.com

Harish Sharma
SCIT
Manipal University Jaipur
Jaipur
India
E-mail: harish27j@gmail.com

Aditya Singh
School of Civil Engineering
Lovely Professional University
Phagwara
India
E-mail: aditya7777777k@gmail.com

Invinder Paul Singh
Department of Agricultural Economics
SKRAU
Bikaner
India

Sarvesh Tanwar
Amity Institute of Information Technology
Amity University Uttar Pradesh
Noida
India
E-mail: s.tanwar1521@gmail.com

Jatin Vaishnav
Department of CSE
School of Engineering and Technology,
CMR University
Bengaluru
India

Aditya Singh Verma
SCIT
Manipal University Jaipur
Jaipur
India
E-mail: asverma314@gmail.com

Arihant Singh Verma
SCIT
Manipal University Jaipur
Jaipur
India
E-mail: arihantsinghverma@gmail.com

Sourabh Singh Verma
SCIT
Manipal University Jaipur
Jaipur
India
E-mail: ssverma80@gmail.com

Gunseerat Kaur

1 Supplementing the Markerless AR with machine learning: Methods and approaches

Abstract: Augmented reality enhances real-world entities with the addition of relevant digital information. This interactive and engaging technology is incorporating itself into wider areas of technical applications. Although it is still in its evolving stage, it has garnered positive responses from audiences who interact with AR-based systems. Industries like gaming, healthcare, and education are benefiting from various augmented reality implementations. The main focus of using AR is to produce more effective interaction with the physical real world as a template and render a separate scenario above it. Initially, a marker-based system was created to project information from a marked region or image with specific characteristics, leading to multiple marketing and retail services adopting this feature. In these applications, a particular marker is sorted through view and detected to project the AR-related information, which further triggers the application' response towards the marker. The baseline strategy includes image capture, processing for markers, tracking the results, and final rendering. However, to create more innovative forms of augmented reality, this concept was merged with machine learning to improve on the aspects that do not revolve around only particular markers. Markerless AR features superimposing graphics on the basis of locations, contour, or projection. This led to the exploration of various machine learning-based algorithms to improve the decision-making accuracy for AR applications. The methodology includes understanding the real environment and fabricating designs accordingly without the need to look for triggering markers. This chapter enlists various markerless AR techniques that use machine learning methods to increase the capability and coherence. The focus is to identify approaches used from machine learning algorithms to amplify efficacy of augmented reality applications using markerless services.

1.1 Introduction

As humans we are likely learning much from our surrounding environment, and our senses work together to procure us information that is processed, memorized, and stored in our minds. It is this highly valuable practice that initiates and creates a channel to extend our knowledge and zeal for learning or noticing the external environment. Augmented Reality changes this practice slightly by altering the method of interaction with the environment, converting conventional methods to superficial ones. It is this

Gunseerat Kaur, Department of Computer Science and Engineering, Lovely Professional University, Punjab, India, e-mail: Gunseerat.22157@lpu.co.in

https://doi.org/10.1515/9783110785234-001

technical enhancement that has amended our perception of vicinity. The implementation of Augmented Reality involves the creation of a specious facade over the real world, thereby creating a separate vision for the actual environment. It creates an approximate or qualitative vision for the surroundings that often shows altered or customized perspectives. The idea of AR is to enhance its user's point of view towards reality.

AR was initiated back in 1968 when Ian Sutherland created the first HMD (Head-mounted display) headset to increase the sensory perception of the world [1]. Since the 1970s, pivotal growth has been observed in establishing augmented reality as an effective technological establishment in a wide variety of applications. Several attempts have been made to incorporate the ideas of using AR into medicine, marketing, and education. Platforms using Augmented Reality have seen huge increments in users for reasons including growing number of internet users, availability of high-quality handheld devices, and less usage of extra equipment that is necessary. Any successful technical enhancement can grow within the user's community when it is readily available to its end-users without any extra hardware installation requirements [2]. AR effects generation generally requires the device to be capable of capturing images of surroundings and processing them, which can be easily achieved through the installation of a plugin or application to superimpose upon objects. Neural networks are providing diverse applications used for enhancing the subjectivity of AR-based projections [3–5]. Since the inception of AR in 1960s, the craze for this technology was enhanced mainly through gaming applications. The first application launched back in 2008, for an advertisement but gradually, as Figure 1.1 shows, the diversity and growth of AR during past two decades have provided a medium to enhance curiosity among users.

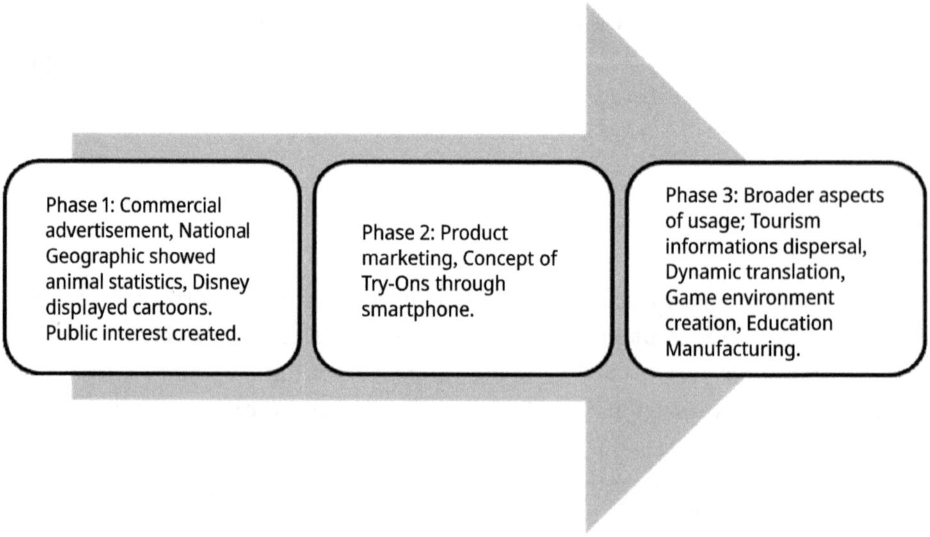

Figure 1.1: Mainstreaming of Augmented reality [6].

This chapter introduces the Augmented Reality as a technology, followed by Section 1.2 which discusses the types of AR, with a focus on markerless. Section 1.3 highlights the relevance of machine learning in AR implementation, while Section 1.4 discusses the contributions of other scholars in this area. Section 1.5 provides an in-depth methodology to show where machine learning is used. Furthermore Section 1.6 provides insights into current case studies in AR based on machine learning technologies, and Section 1.7 presents the conclusion.

1.2 Motivation

Augmented Reality has surely piqued the curiosity of many researchers and has played an important role in creation of innovative practices involving it. Machine learning is a pivotal field that is continuously enhancing its properties and increasing its boundaries. There is a growing amount of literature that supports the benefits of incorporating these two powerful technologies, as the strength of AR can be improved and enhanced through well-developed and practiced methods of ML. Researching through these papers, the results suggest considerable efforts of joining these technologies. This chapter aims to shed light on some of these pivotal studies while investigating and understanding the intricate balance between AR and ML. One major challenge that poses a current research gap in this approach lies within the devices projecting AR models. As multiple instances use common smartphones that have limited resources for computation, issues such as latency, speed, and dynamics of rendering AR models are delayed due to longer computations. To date, only a few studies have investigated this aspect as well.

1.3 How have AI and blockchain transformed AR?

Classifying AR into various different categories depends on the type of hardware used or how the end result is projected. In accordance with the theme of this chapter, the classification paradigm is the basis of how AR understands the real world and projects the final rendering [7].

As Figure 1.2 shows, the main classification at this level is between Marked AR and Markerless AR. Marker-based AR is an approach where a target object is identified, and a pivot is used to situate the rendered object on that specific surface.

This is an initial approach that is generally found in smartphone cameras to project different filters. Identification of a point is relevant in this form of AR. [8] have compared on how CNN and SVM can be used to identify the spot to render AR, showing 96.7 % and 92.5 % respectively. Their application is designed to recognize and project

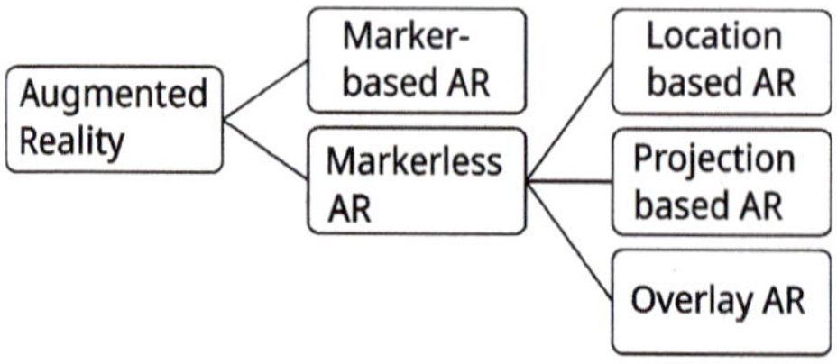

Figure 1.2: Types of augmented reality.

Figure 1.3: A marker-based approach requires markers such as QR codes to trigger an event. Image credit: [9].

the alphabets for children in a learning app. Figure 1.3 shows a person making a digital purchase through a QR-code.

Marker-based AR is successful when the point of reference is well-defined and output is based around it. However, in situations where the marker is not detected it proves to be unfruitful [2]. This led to creation of Markerless AR, which relies on incoming data from various sensors to place a rendered output in context with the real world. Mainly, as listed in Table 1.1, a simultaneous localization and mapping criteria are used, which creates an estimate to place objects in the real world.

Table 1.1: Comparisons between marker-less and marker-based AR [2, 8, 10].

Comparisons	Marker-based AR	Marker-less AR
Reference point	Needed	Not needed
Image Description	Pre-provided	Calculated
Sensors	Accelerometer not needed	Accelerometer, Gyroscope needed
Recognition of planar objects	Limited	Not limited
Lighting adaptability	Robust, Limited	Adjustable to increasing and decreasing light

1.4 Where machine learning meets Augmented Reality

The functional process of AR-based model creation includes a flow process that captures, analyzes, and iterates over the given paradigm to create a perpetual AR object in the environment. Figure 1.4 shows how the process starts at the first stage by capturing the image from either smartphone cameras or individual cameras.

These images are provided to computer vision which then identifies and classifies the various objective and structural specifications for the captured object. An AR model can be created very easily but the challenges faced are mainly in terms of calculating light, so that it seems real with respect to lighting in the real world [12]. Placement of the

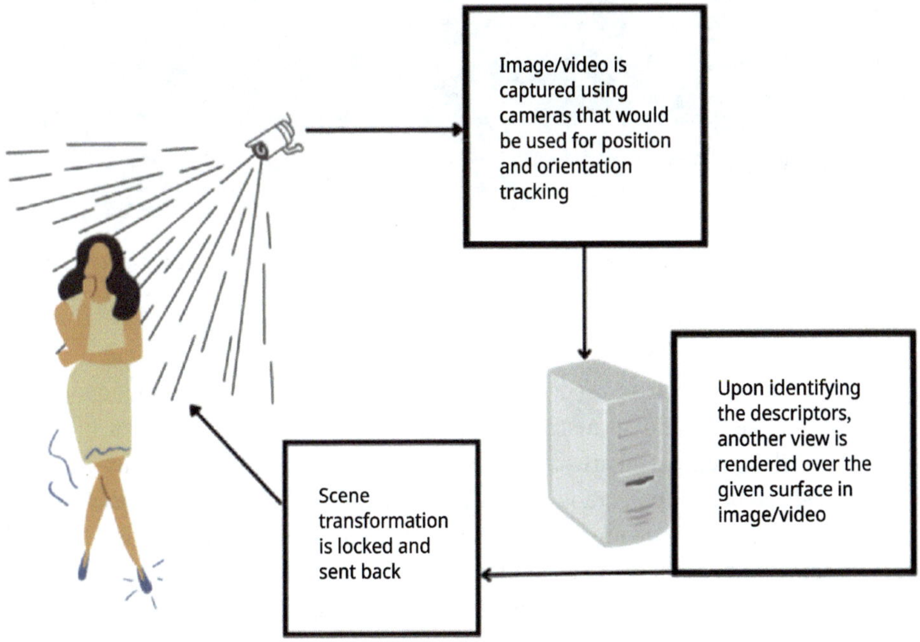

Figure 1.4: Object pose tracking and prediction to render AR-based projection for customer to choose [11].

model and positioning it correctly in terms of size and scale also make a huge difference and can be challenging at times. Adaptation in augmentation of solid objects is enhanced if context awareness can be incremented using different strategies to elaborate on the features of the surrounding environment. This would ease the practice of redesigning and modifying the AR model accordingly. After considering all these options, a user display is constructed to provide the final model in the real world [10]. Through this process at various stages, machine learning plays an important role in tracking features, creating checkpoints, estimation of light sources, and setting pivots. These mechanisms are enhanced in their usage using algorithms based on neural networks. Machine Learning enhances the functionality of AR. Initially, when an image is captured via a camera, machine learning is used to keep track of objects in that image.

Figure 1.5: Contour-based AR application that shows checkpoints and delineates road points for its user [13]. Image credit: [14].

A feature-based approach is used to detect planar surfaces to place models in the real world. The convolutional neural network architecture AlexNet is used to create Deep AR by [3], which performs well for 2D image generation using feature localization approach. Another application of machine learning techniques being used in conjunction with AR is real-time text translation. The main motive is to provide fast and correct results. Proper text detection and translation using natural language processing presents an effective approach [15]. Using machine learning to enhance the functionality of AR creates a wide variety of scenarios of interest such as object recognition, detection, object labeling, text and speech recognition as depicted in Figure 1.5.

1.5 Related work

Markerless AR systems use patterns, colors or light sources to identify the surroundings and provide better user experience. [16] authors have remarkably created a low-cost IR imaging-based AR setup that can be attached to the windshield to highlight obstacles while night driving. They have shown usage of SVM classifiers to classify harmful and harmless objects in low light for the driver. Similarly, [17] have worked with SVM and Long short-term memory techniques that can easily classify the gestures for creating a guiding system. SVM has better hold on smaller samples which can handle nonlinear patterns. Another example by [18] shows usage of structured linear SVM to keep track of frames, rotation and motion of objects in images and apply AR objects over their surface. They used 10 frames to feed and understand surroundings before applying AR to recognize and occlude AR objects on real-time surfaces.

[19] have shifted towards using another efficient algorithm, namely Convolutional Neural Networks (CNN), preferred for its smaller size and efficient processing time. An AR-based inspection model is created which recognizes faulty or non-faulty machinery in a production unit. Implementing a sub-category in CNN, they have sorted to gather 70 % accuracy on a non-marker-based schema to identify machinery and identify its faults. Localization plays an important role in markerless AR, therefore, [20] have also focused on adding an extra semantic layer made of CNN to their advantage in understanding geometry and real-world concepts. They have created a system to recognize labels on machines, knob and valve values, pressure information and classify them from their normal to abnormal stages to generate alarms.

Based on a similar principle of execution, [17] have shown how gesture recognition can be enhanced to facilitate AR. Their study shows construction of an assistance system that would produce information for objects that are being pointed at. Images are captured through device cameras, which are passed towards feature extractions and fed through mask R-CNN that identifies and forwards data towards AR projecting devices; the technique shows an accuracy of 79 % while identifying objects.

As an object localization algorithm CNN shows very promising results. [21] have worked with chess piece recognition using contour-based chamfer matching and compared it against CNN. Upon observation, they concluded the chamfer matching also shows similar results but is lightweight as compared to CNN, presenting another front for exploration.

[22] have surfaced a method using CNN to detect cable brackets in aircraft, they detect and superimpose missing brackets and their probable structure over the images to provide insights for users. For markerless AR, localization is pivotal, hence, [23] have worked on improving the learning curve for mobile-based AR applications to process data at edge nodes, using image input with a smaller size can directly affect the algorithms learning curve.

Vast studies have used CNN due to its lesser pre-processing requirements. Although, image classification algorithms like Naive Bayes have also been used in multiple applications such as tree age prediction, wherein AR module collects the dimensions and other features of the tree feeding them into the Naive Bayes classifier that will predict the age. Similarly, [24] has shown usage of Naive Bayes classifier to fetch details of augmented maps and detect damaged roads. Scene Recognition shows several applications of Naive Bayes in conjunction with AR [25]. Numerous studies have also shown the usage of KNN to create tools like AR-based piano learning systems [26], face detection and information presenter [27], or accommodation finder based on AR mapping or template tracking systems.

1.6 Methodology

In markerless AR systems, there are four categories, as shown in Figure 1.2. The idea is to use a certain parameter that can compensate for the lack of a pivot point to render AR objects. Three categories fall under markerless AR: location-based, projection-based, and overlay AR.

1.6.1 Location-based AR

Recent advances in markerless AR have led to effective and precise location and position tracking, which can easily set up a pretext for placing rendered AR models. Data from GPS, accelerometers, gyroscopes, and other such sensors are sent for processing information about location and orientation with respect to the Earth's magnetic field. Figure 1.6 illustrates the process of information gathering for location-based processing.

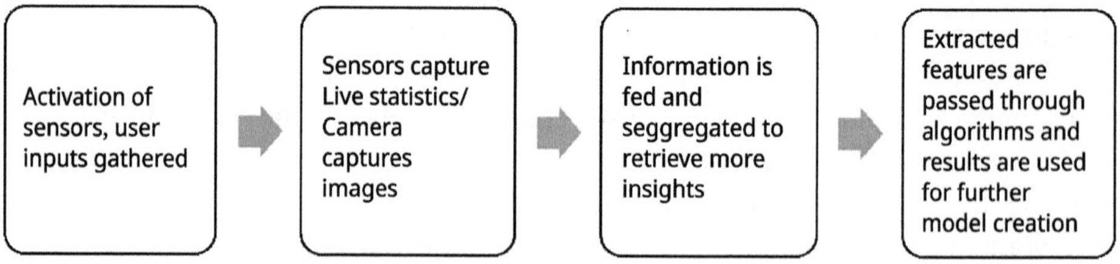

Figure 1.6: Process of information gathering [28].

The amount of physical motion or sedentary state can be predicted using algorithms such as linear regression or ROC curve [29]. For outdoor locations, establishing AR models is easier as GPS signals are not hindered and are precise. However, inconsistencies occur in indoor systems for location-based AR systems as the data locations and paths are often overlapped or not proper [30]. [31] has shown the usage of AR with SVM and

Figure 1.7: Usage of location-based AR to display information on local tourist sites [33]. Image credit: [9].

MLP to create a robotic system that can adapt and map the indoor paths in a building, creating an estimated map to provide signals and route information for the creation of an AR-based visual positioning system. Another approach incorporates Wi-Fi signal addition to GPS signals to curb the challenge of location precision in indoor systems for AR modeling [32]. Figure 1.7 is illustrating the usage of location based AR to display

Accuracy plays a pivotal role in location-based AR, as it determines the altitude and position of the rendered object. For a better viewing experience, the distance between the viewing device and projected model needs to be calculated to achieve impressive results. Applications for location-based AR include game development to provide players with a real-life experience through their handheld devices. Following the interest of users in playing AR-based games, marketing and advertisers have started showcasing products with AR, and displaying computer-aided models in the field of interior and exterior designing has garnered attention. Education systems have also incorporated the strategy of using location-based AR in creating interactive lessons for delivering pedagogy.

1.6.2 Projection-based AR

Projection-based AR, also known as spatial Augmented Reality, is a technique used for creating shared AR experiences by projecting the rendered model over an area. It does

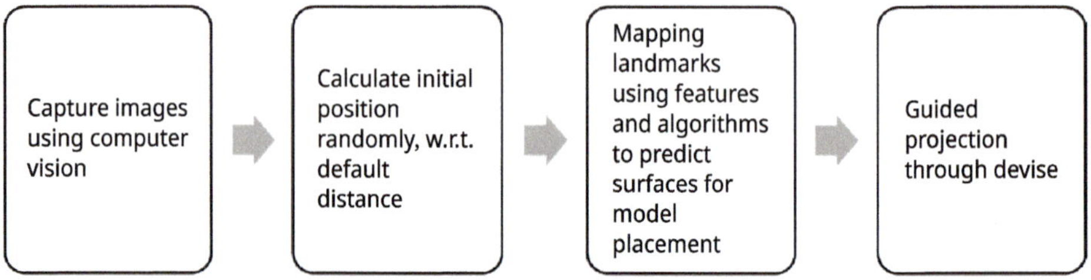

Figure 1.8: Flow process for projection-based AR [23, 34].

not use devices for viewing AR, but instead projects them directly into real-world space, providing an insight into mixed reality. To achieve this, multi-purpose cameras are needed to keep track of depth, lighting, and occlusion [12]. Figure 1.8 shows a process flow that starts with capturing information about the environment using computer vision and applying machine learning-based algorithms to enhance the process of identifying and seeking the coordinates to render an effective projection over the surface.

An application to detect poses for an object in 3D space was created by [11], who used a Random Ferns classifier to obtain the classification of poses using infrared images, which reduced the delay in projection [4]. [35] has used an approach to track and project a user's interaction with a smart home system. Deep learning and computer vision are used to track activity for a user by classifying the actions towards different planes of action, namely nearby planes, pointing planes, and looking planes. In healthcare systems, [25] has shown the usage of effective visualization techniques using AR. The model keeps track of C-arm and Kinect's depth sensor alongside the patient's anatomy to help surgeons understand the surgical scene in the context of X-ray data. Similarly, [34] has

Figure 1.9: A customer using head-mounted display to choose between various items [36]. Image credit: [37].

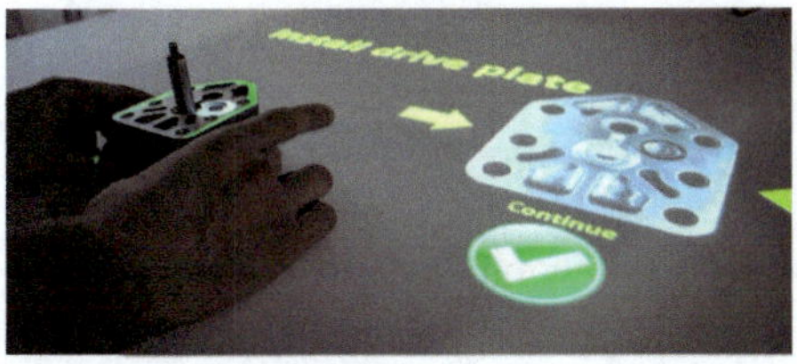

Figure 1.10: Lightguidesys [40] has created guided systems to help employees understand manufacturing processes. Image credit: [40].

shown a projector-based system that can outline the tumor boundaries in laparoscopic partial nephrectomy, aiding in precise dissections as shown in Figure 1.9 and 1.10.

Combining healthcare and education, [38] has used decision trees to keep track of trainees' learning on AR surgical simulators. Furthermore, the process of projection can be used in displaying useful information. [39] has used CNN, ANN, and SVM to predict and compare the motion on the road and to drive the vehicle. An HMD is also used to collect data at regular intervals for processing, and the state of the vehicle at any point is projected for its passengers. Applications for projected AR also include manufacturing processes, where employees are given instructions or steps to follow, reducing mistakes.

1.6.3 Overlay AR

Overlay AR, also called superimposition AR, superimposes the rendered model over the current target, either fully or partially, depending on the equipment used in viewing. It allows the user to see an alternative view of the real world in a separate space. Unlike projection-based systems, an overlay AR might require extra hardware to view the simulations constructed. [41] has created a driver information system HUD (Head-up display), which acts as an active vehicle safety system. It uses VANETs to interact with corresponding nodes to capture information about traffic congestion. Road objects are detected using R-CNN with less delay in data processing at each node. Superimposed AR can also be used to check and imply default settings in manufacturing units. [30] has worked on a system that uses MobileNets, a model based on CNNs. This allows the detection of values on indoor installed industrial equipment and compares them against normal values. In case of discrepancy, it superimposes the new required values over the current ones to signify changes. Ikea [42], a furniture-based organization, has promoted self-designing applications for interiors. These applications allow users to view a superimposed layout of new products in the original dimensions.

1.7 Case studies: where Augmented Reality needs machine learning?

Augmented Reality deciphers information from its environment and hence requires an understanding of this environment in great detail. Learning about its surroundings plays a vital role in determining the type of augmentation that can be created and dynamically enhanced based on its supplied data – data that will be recorded simultaneously while generating the AR object or before placing the object. Learning about these roles in detail will provide insight into how machine learning algorithms aid augmentations. Figure 1.11 enlists the variety of purposes and methods that are based on machine learning but also allow Augmented Reality applications to enhance their functionality. For instance, social media uses a variety of filters and theme-based camera-based objects that allow the user to modify and redefine images [43].

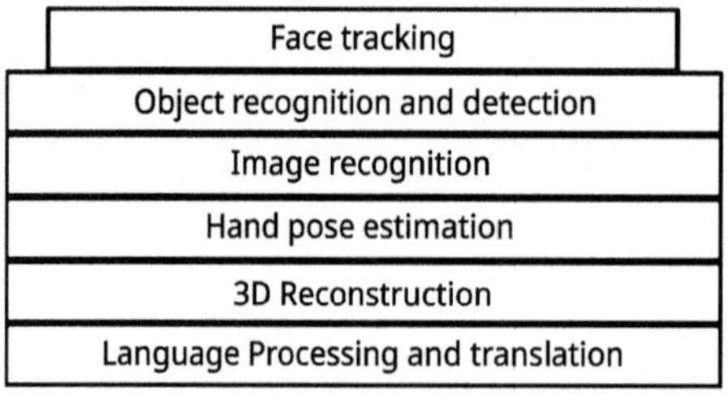

Figure 1.11: Various fields from machine learning used to enhance Augmented Reality [44].

It is objectively important to understand the methods that allow markerless AR to succeed in working with different paradigms from machine learning used in creating a base for AR objects to be placed. Figure 1.11 shows six very effective concepts that are molded to provide a base for AR. Often, face tracking creates a clear platform to identify and track the placement of AR objects with respect to the face, such as filters and animations. A point distribution model [45] determines the coordinate points that are used as identifiable features and creates a baseline for understanding the shape in 3D. Methods like Procrustes analysis are used to determine the transformation, such as rotation, scaling, and superimpositions of objects. These features are recently released as a 3D face mesh estimation tool based on 468 facial landmarks that determine the facial pose and apply a variety of textures [46].

In mobile computing, AR has grabbed attention at the forefront, and an issue that arises here is while recording videos using AR, often the accuracy is lowered, and an apparent latency is also observed at the user's end. [47] have proposed VCMaker, which is a combination of reinforcement learning to select the configurations while recording the video and further enables a neural network model that is responsible for selecting and predicting future object placement. It has shown a reduction in energy consumption by

25.2 %–45.7 % with an increase of 20–32 % in detection accuracy. Accuracy and latency have been addressed through a content-aware encoding system. These systems contain convolutional networks that allow for the detection of similarities and considerations of classification. The issue with AR projection through the markerless method that has been concluded is that neural inference and intensive calculations are limited in a handheld device. Therefore, it causes latency and accuracy to be directly proportional to the user's choice of configuration for a particular device.

Table 1.2 collects some unique methods and cases where AR has shown an intriguing approach towards problem-solving in daily real-life scenarios, in education, manufacturing, or logistics. The opportunities for applying AR in different means are explored extensively.

Table 1.2: Machine learning in Augmented Reality.

Authors/ Contributors	AR usage	Machine learning aspect	Conclusions
[45]	Spectacle frame selection/projection on users' face dependent on image input	Face tracking, Pose estimation	AR needs correct algorithm to feature better accuracy while processing objects in real-time.
[48]	AR-based navigation system to guide the placement of needles to stimulate sacral nerve	Optical tracking system	AR needs better depth perception while computing the trajectory of needle puncture direction.
[49]	Internal logistics mapped and controlled by simultaneous localization and mapping in real factory layout	Localization, Mapping	AR is used for visualizing the 3D objects in real factory layout to facilitate seamless planning.
[50]	AR-based product configuration on a fire truck	Model recognition and tracking	AR helps the firemen gather tools/equipment by locating them from the truck instead of following the complete product catalogue, all information can be gathered through viewing AR objects and their subparts in 3D.
[51]	AR-based timber drilling and cutting methodology	Object tracking	A markerless technique in conjunction with SLAM and depth tracking for drilling holes and providing guidance to users.

Table 1.2 (continued)

Authors/ Contributors	AR usage	Machine learning aspect	Conclusions
[52]	AR-based shopping assistance	Object recognition, tracking	AR-based object recognition while in-store purchases are enhanced. GPS tracking to enhance users' tracking and accordingly suggest in-house and online discounts/entities.
[53]	AR-based assessment evaluation	Sentiment analysis, Natural language processing, Recurring neural networks	Skill assessment of an operator, through the means of AR projection to help understand test takers' proficiency in a simulated environment.
[54]	AR-based markerless construction modelling	Object verification, tracking	Markerless approach for a larger section of building management.
[55]	AR based indoor virtual landmark system	Sensor and vision tracking, Map/3D construction	A hybrid markerless tracking system for indoor objects, provision for virtual displays of artifacts, equipment, models.
[56]	AR-based disaster management training system	3D reconstruction, Object tracking, Movement tracking	3D construction of scenarios based on markerless environment to simulate disasters and help learners avoid and calibrate themselves accordingly.
[57]	Algorithm for Markerless AR prototype	Object tracking/recognition with SVM, Image recognition	A speedy, low latency Markerless AR system prototype.
[58]	AR-based application for divers	Sound localization, Computer Vision, Neural networks	AR for understanding deep sea life and artifacts and their history through AR projections while diving.
[59]	AR-based sports training	Hand pose estimation, movement tracking	Reduction of latency while applying AR in sports training can provide multitude of opportunities for players
[60]	AR-RFID based system for logistics	Object tracking, Movement tracking	Markerless AR to identify GPS-enabled locations in frames to determine locations for usability in logistics.

1.8 Conclusion

Overall, machine learning algorithms are crucial in the markerless AR context, where the AR model needs to be placed without the need for a physical marker. Machine learning techniques like face tracking, pose estimation, localization, mapping, and object recognition are used to create a baseline for understanding the real-world environment and enabling dynamic placement of AR models. Several case studies demonstrate the potential of AR in various fields, including education, manufacturing, logistics, and sports training, where AR-based applications have been developed using machine learning techniques like object tracking, hand pose estimation, and sentiment analysis. With the growing number of internet users and high-quality handheld devices, the availability of AR effects generation can be readily achieved through installation of plugins or applications without the need for extra hardware installation requirements. CNN algorithms have shown promising results in reducing delays in dynamic processing, making them suitable for AR applications. Therefore, machine learning algorithms have a significant role to play in creating and rendering AR models in the markerless AR context.

Bibliography

[1] R. Behringer, G. Klinker, and D. Mizell, "International Workshop on Augmented Reality 1998—Overview and Summary," In IWAR '98 Proceedings of the International Workshop on Augmented Reality: Placing Artificial Objects in Real Scenes: Placing Artificial Objects in Real Scenes, p. International Workshop on Augmented Reality 1998—o, 1998.
[2] P. Q. Brito and J. Stoyanova, "Marker versus Markerless Augmented Reality. Which Has More Impact on Users?," International Journal of Human-Computer Interaction, vol. 34, no. 9, pp. 819–833, Sep. 2017, https://doi.org/10.1080/10447318.2017.1393974.
[3] O. Akgul, H. I. Penekli, and Y. Genc, "Applying Deep Learning in Augmented Reality Tracking," In Proceedings—12th International Conference on Signal Image Technology and Internet-Based Systems, SITIS 2016, pp. 47–54, Apr. 2017, https://doi.org/10.1109/SITIS.2016.17.
[4] N. Hashimoto and D. Kobayashi, "Dynamic Spatial Augmented Reality With a Single IR Camera," In SIGGRAPH 2016—ACM SIGGRAPH 2016 Posters, Jul. 2016, https://doi.org/10.1145/2945078.2945083.
[5] S. S. P. Gannamani, P. Srivani, S. Adarsh, V. Shashank, and M. Bharath Shreya, "Tree Age Predictor Using Augmented Reality and Image Processing Techniques," In CSITSS 2021—2021 5th International Conference on Computational Systems and Information Technology for Sustainable Solutions, Proceedings, 2021, https://doi.org/10.1109/CSITSS54238.2021.9683736.
[6] "The Mainstreaming of Augmented Reality: A Brief History." https://hbr.org/2016/10/the-mainstreaming-of-augmented-reality-a-brief-history (accessed Jul. 28, 2022).
[7] J. Peddie, "Types of Augmented Reality," In Augmented Reality pp. 29–46, 2017, https://doi.org/10.1007/978-3-319-54502-8_2.
[8] J. C. P. Cheng, K. Chen, and W. Chen, "Comparison of Marker-Based and Markerless AR: A Case Study of An Indoor Decoration System," In Lean & Computing in Construction Congress (LC3), pp. 483–490, Jul. 2017, https://doi.org/10.24928/JC3-2017/0231.
[9] "person holding white ipad with black case photo – Free Image on Unsplash." https://unsplash.com/photos/CyX3ZAti5DA (accessed Jul. 28, 2022).

[10] A. K. Dash, S. K. Behera, D. P. Dogra, and P. P. Roy, “Designing of Marker-Based Augmented Reality Learning Environment for Kids Using Convolutional Neural Network Architecture,” Displays, vol. 55, pp. 46–54, Dec. 2018, https://doi.org/10.1016/J.DISPLA.2018.10.003.

[11] H. Ro, J. H. Byun, Y. J. Park, and T. D. Han, “Display Methods of Projection Augmented Reality Based on Deep Learning Pose Estimation,” In ACM SIGGRAPH 2019 Posters, SIGGRAPH 2019, Jul. 2019, https://doi.org/10.1145/3306214.3338608.

[12] A. Holynski and J. Kopf, “Fast Depth Densification for Occlusion-Aware Augmented Reality,” ACM Transactions on Graphics, Dec. 2018, https://doi.org/10.1145/3272127.3275083.

[13] “Lining up AR Features While Weaving Through Traffic With the Vision SDK.” https://www.mapbox.com/blog/lining-up-ar-features-while-weaving-through-traffic-with-the-vision-sdk (accessed Apr. 16, 2022).

[14] “Lining up AR Features While Weaving Through Traffic With the Vision SDK | by Mapbox | Maps for Developers.” https://blog.mapbox.com/lining-up-ar-features-while-weaving-through-traffic-with-the-vision-sdk-661c28363da4 (accessed Jul. 28, 2022).

[15] “Cloud Translation | Google Cloud.” https://cloud.google.com/translate/ (accessed Apr. 07, 2022).

[16] K. S. Ramasubramaniam and R. Bhat, “LcAR-Low Cost Augmented Reality for the Automotive Industry,” In 2018 IEEE International Conference on Consumer Electronics, ICCE 2018, vol. 2018-January, pp. 1–3, Mar. 2018, https://doi.org/10.1109/ICCE.2018.8326234.

[17] T. Wang, X. Cai, L. Wang, and H. Tian, “Interactive Design of 3D Dynamic Gesture Based on SVM-LSTM Model,” International Journal of Mobile Human Computer Interaction (IJMHCI), vol. 10, no. 3, pp. 49–63, Jan. 1AD 2018, https://doi.org/10.4018/IJMHCI.2018070104, https://services.igi-global.com/resolvedoi/resolve.aspx?doi=10.4018/IJMHCI.2018070104.

[18] Z. W. Gui, “Register Based on Efficient Scene Learning and Keypoint Matching for Augmented Reality System,” In 2016 International Conference on Image, Vision and Computing, ICIVC 2016, pp. 79–85, Sep. 2016, https://doi.org/10.1109/ICIVC.2016.7571277.

[19] T. Perdpunya, S. Nuchitprasitchai, and P. Boonrawd, “Augmented Reality with Mask R-CNN (ARR-CNN) inspection for Intelligent Manufacturing,” ACM International Conference Proceeding Series, Jun. 2021, https://doi.org/10.1145/3468784.3468788.

[20] J. Izquierdo-Domenech, J. Linares-Pellicer, and J. Orta-Lopez, “Supporting Interaction in Augmented Reality Assisted Industrial Processes Using a CNN-Based Semantic Layer,” In Proceedings—2020 IEEE International Conference on Artificial Intelligence and Virtual Reality, AIVR 2020, pp. 27–32, Dec. 2020, https://doi.org/10.1109/AIVR50618.2020.00014.

[21] Y. Xie, G. Tang, and W. Hoff, “Chess Piece Recognition Using Oriented Chamfer Matching with a Comparison to CNN,” In Proceedings – 2018 IEEE Winter Conference on Applications of Computer Vision, WACV 2018, vol. 2018-January, pp. 2001–2009, May 2018, https://doi.org/10.1109/WACV.2018.00221.

[22] G. Zhao, J. Hu, W. Xiao, and J. Zou, “A Mask R-CNN Based Method for Inspecting Cable Brackets in Aircraft,” Chinese Journal of Aeronautics, vol. 34, no. 12, pp. 214–226, Dec. 2021, https://doi.org/10.1016/J.CJA.2020.09.024.

[23] B. G. Mark, E. Rauch, and D. T. Matt, “Study of the Impact of Projection-Based Assistance Systems for Improving the Learning Curve in Assembly Processes,” Procedia CIRP, vol. 88, pp. 98–103, Jan. 2020, https://doi.org/10.1016/J.PROCIR.2020.05.018.

[24] A. A. Rafique, A. Jalal, and A. Ahmed, “Scene Understanding and Recognition: Statistical Segmented Model using Geometrical Features and Gaussian Naïve Bayes,” 2019 International Conference on Applied and Engineering Mathematics (ICAEM) https://doi.org/10.1109/ICAEM.2019.8853721.

[25] O. Pauly, B. Diotte, P. Fallavollita, S. Weidert, E. Euler, and N. Navab, “Machine Learning-Based Augmented Reality for Improved Surgical Scene Understanding,” Computerized Medical Imaging and Graphics, vol. 41, pp. 55–60, Apr. 2015, https://doi.org/10.1016/J.COMPMEDIMAG.2014.06.007.

[26] H. Zeng, X. He, and H. Pan, "A New Practice Method Based on KNN Model to Improve User Experience for an AR Piano Learning System," In Virtual, Augmented and Mixed Reality. Applications and Case Studies, Lecture Notes in Computer Science (including subseries Lecture Notes in Artificial Intelligence and Lecture Notes in Bioinformatics), vol. 11575 LNCS, pp. 398–409, 2019, https://doi.org/10.1007/978-3-030-21565-1_27.

[27] A. Golnari, H. Khosravi, and S. Sanei, "DeepFaceAR: Deep Face Recognition and Displaying Personal Information via Augmented Reality," In Iranian Conference on Machine Vision and Image Processing, MVIP, vol. 2020–February, Feb. 2020, https://doi.org/10.1109/MVIP49855.2020.9116873.

[28] W. van Woensel, P. C. Roy, S. S. R. Abidi, and S. R. Abidi, "Indoor Location Identification of Patients for Directing Virtual Care: An AI Approach Using Machine Learning and Knowledge-Based Methods," Artificial Intelligence in Medicine, vol. 108, p. 101931, Aug. 2020, https://doi.org/10.1016/J.ARTMED.2020.101931.

[29] V. Farrahi, M. Niemelä, M. Kangas, R. Korpelainen, and T. Jämsä, "Calibration and Validation of Accelerometer-Based Activity Monitors: A Systematic Review of Machine-Learning Approaches," Gait & Posture, vol. 68, pp. 285–299, Feb. 2019, https://doi.org/10.1016/J.GAITPOST.2018.12.003.

[30] H. Subakti and J. R. Jiang, "Indoor Augmented Reality Using Deep Learning for Industry 4.0 Smart Factories," In Proceedings – International Computer Software and Applications Conference, vol. 2, pp. 63–68, Jun. 2018, https://doi.org/10.1109/COMPSAC.2018.10204.

[31] A. C. Seckin, "Adaptive Positioning System Design Using AR Markers and Machine Learning for Mobile Robot," In 5th International Conference on Computer Science and Engineering, UBMK 2020, pp. 160–164, Sep. 2020, https://doi.org/10.1109/UBMK50275.2020.9219475.

[32] G. K. Kamalam, S. Joshi, M. Maheshwari, K. Senthamil Selvan, S. Shaukat Jamal, S. Vairaprakash, M. Alhassan, et al., "Augmented Reality-Centered Position Navigation for Wearable Devices with Machine Learning Techniques," Journal of Healthcare Engineering, vol. 2022, no. Special Issue, 2022. Accessed: Apr. 16, 2022. [Online]. Available: https://www.hindawi.com/journals/jhe/2022/1083978/.

[33] P. Amirian and A. Basiri, "Landmark-Based Pedestrian Navigation Using Augmented Reality and Machine Learning," In Lecture Notes in Geoinformation and Cartography, pp. 451–465, 2016, https://doi.org/10.1007/978-3-319-19602-2_27.

[34] P. Edgcumbe, R. Singla, P. Pratt, C. Schneider, C. Nguan, and R. Rohling, "Follow the Light: Projector-Based Augmented Reality Intracorporeal System for Laparoscopic Surgery," Journal of Medical Imaging, vol. 5, no. 2, p. 021216, Feb. 2018, https://doi.org/10.1117/1.JMI.5.2.021216.

[35] "A Study on Augmented Reality-based Positioning Service Using Machine Learning -Proceedings of the Korean Institute of Information and Commucation Sciences Conference | Korea Science." https://www.koreascience.or.kr/article/CFKO201714956117081.page (accessed Apr. 16, 2022).

[36] "Rethinking Interface Assumptions in AR: Selecting Objects | by Aaron Cammarata | Google Play Apps & Games | Medium." https://medium.com/googleplaydev/rethinking-interface-assumptions-in-ar-selecting-objects-a6675c7c1d1c (accessed Apr. 16, 2022).

[37] "Woman Technology Science – Free photo on Pixabay." https://pixabay.com/photos/woman-technology-science-design-6929333/ (accessed Jul. 28, 2022).

[38] H. Ghandorh, "Prediction of Users' Performance in Surgical Augmented Reality Simulation-Based Training Using Machine Learning Techniques," In Digital Future of Healthcare, pp. 95–108, Nov. 2021, https://doi.org/10.1201/9781003198796-6.

[39] S. Murugan, A. Sampathkumar, S. Kanaga Suba Raja, S. Ramesh, R. Manikandan, and D. Gupta, "Autonomous Vehicle Assisted by Heads up Display (HUD) with Augmented Reality Based on Machine Learning Techniques," Studies in Systems, Decision and Control, vol. 412, pp. 45–64, 2022, https://doi.org/10.1007/978-3-030-94102-4_3.

[40] "6 Uses of Augmented Reality for Manufacturing In Every Industry—LightGuide." https://www.lightguidesys.com/6-uses-of-augmented-reality-for-manufacturing-in-every-industry/ (accessed Apr. 16, 2022).

[41] L. Abdi and A. Meddeb, "Driver Information System: A Combination of Augmented Reality, Deep Learning and Vehicular Ad-Hoc Networks," Multimedia Tools and Applications, vol. 77, no. 12, pp. 14673–14703, Jun. 2018, https://doi.org/10.1007/S11042-017-5054-6/FIGURES/16.

[42] “Design your Room—IKEA.” https://www.ikea.com/in/en/planners/design-your-dream-home-pub66945dd9 (accessed Apr. 16, 2022).
[43] C. Flavián, S. Ibáñez-Sánchez, and C. Orús, “User Responses Towards Augmented Reality Face Filters: Implications for Social Media and Brands,” pp. 29–42, 2021, https://doi.org/10.1007/978-3-030-68086-2_3.
[44] “10 Ways Augmented Reality Uses Machine Learning.” https://connectjaya.com/10-ways-augmented-reality-uses-machine-learning/ (accessed Jul. 28, 2022).
[45] E. Setyati, D. Alexandre, and D. Widjaja, “Face Tracking Implementation with Pose Estimation Algorithm in Augmented Reality Technology,” Procedia – Social and Behavioral Sciences, vol. 57, pp. 215–222, Oct. 2012, https://doi.org/10.1016/J.SBSPRO.2012.09.1177.
[46] “Google Developers Blog: MediaPipe 3D Face Transform.” https://developers.googleblog.com/2020/09/mediapipe-3d-face-transform.html (accessed Jul. 26, 2022).
[47] N. Chen, S. Zhang, S. Quan, Z. Ma, Z. Qian, and S. Lu, “VCMaker: Content-Aware Configuration Adaptation for Video Streaming and Analysis in Live Augmented Reality,” Computer Networks, vol. 200, p. 108513, Dec. 2021, https://doi.org/10.1016/J.COMNET.2021.108513.
[48] R. Moreta-Martínez, I. Rubio-Pérez, M. García-Sevilla, L. García-Elcano, and J. Pascau, “Evaluation of Optical Tracking and Augmented Reality for Needle Navigation in Sacral Nerve Stimulation,” Computer Methods and Programs in Biomedicine, vol. 224, p. 106991, Sep. 2022, https://doi.org/10.1016/J.CMPB.2022.106991.
[49] A. Rohacz, S. Weißenfels, and S. Strassburger, “Concept for the Comparison of Intralogistics Designs With Real Factory Layout Using Augmented Reality, SLAM and Marker-Based Tracking,” Procedia CIRP, vol. 93, pp. 341–346, Jan. 2020, https://doi.org/10.1016/J.PROCIR.2020.03.039.
[50] F. Bellalouna, “The Augmented Reality Technology as Enabler for the Digitization of Industrial Business Processes: Case Studies,” Procedia CIRP, vol. 98, pp. 400–405, Jan. 2021, https://doi.org/10.1016/J.PROCIR.2021.01.124.
[51] A. Settimi, J. Gamerro, and Y. Weinand, “Augmented-Reality-Assisted Timber Drilling With Smart Retrofitted Tools,” Automation in Construction, vol. 139, p. 104272, Jul. 2022, https://doi.org/10.1016/J.AUTCON.2022.104272.
[52] P. Ashok Kumar and R. Murugavel, “Prospects of Augmented Reality in Physical Stores’s Using Shopping Assistance App,” Procedia Computer Science, vol. 172, pp. 406–411, Jan. 2020, https://doi.org/10.1016/J.PROCS.2020.05.074.
[53] D. Mourtzis, J. Angelopoulos, V. Siatras, and N. Panopoulos, “A Methodology for the Assessment of Operator 4.0 Skills based on Sentiment Analysis and Augmented Reality,” Procedia CIRP, vol. 104, pp. 1668–1673, Jan. 2021, https://doi.org/10.1016/J.PROCIR.2021.11.281.
[54] H. S. Kim, Sangmi-Park, Sunju-Han, and L. S. Kang, “AR-based 4D CAD System Using Marker and Markerless Recognition Method,” Procedia Engineering, vol. 196, pp. 29–35, Jan. 2017, https://doi.org/10.1016/J.PROENG.2017.07.169.
[55] F. N. Afif and A. H. Basori, “Orientation Control for Indoor Virtual Landmarks based on Hybrid-based Markerless Augmented Reality,” Procedia – Social and Behavioral Sciences, vol. 97, pp. 648–655, Nov. 2013, https://doi.org/10.1016/J.SBSPRO.2013.10.284.
[56] H. Mitsuhara, C. Tanimura, J. Nemoto, and M. Shishibori, “Expressing Disaster Situations for Evacuation Training Using Markerless Augmented Reality,” Procedia Computer Science, vol. 192, pp. 2105–2114, Jan. 2021, https://doi.org/10.1016/J.PROCS.2021.08.218.
[57] P. Khandelwal, P. Swarnalatha, N. Bisht, and S. Prabu, “Detection of Features to Track Objects and Segmentation Using GrabCut for Application in Marker-less Augmented Reality,” Procedia Computer Science, vol. 58, pp. 698–705, Jan. 2015, https://doi.org/10.1016/J.PROCS.2015.08.090.
[58] F. Bruno et al., “Underwater Augmented Reality for Improving the Diving Experience in Submerged Archaeological Sites,” Ocean Engineering, vol. 190, p. 106487, Oct. 2019, https://doi.org/10.1016/J.OCEANENG.2019.106487.
[59] P. Soltani and A. H. P. Morice, “Augmented Reality Tools for Sports Education and Training,” Computers and Education, vol. 155, p. 103923, Oct. 2020, https://doi.org/10.1016/J.COMPEDU.2020.103923.
[60] E. Ginters, A. Cirulis, and G. Blums, “Markerless Outdoor AR-RFID Solution for Logistics,” Procedia Computer Science, vol. 25, pp. 80–89, Jan. 2013, https://doi.org/10.1016/J.PROCS.2013.11.010.

Arzoo, Kiranbir Kaur, and Salil Bharany

2 IOT in AR/VR for flight simulation

Abstract: Simulative environments generally provide better clarity in terms of learning compared to command user interfaces. Raster and random scan systems were commonly used in simulative environments. With advancements, Augmented Reality (AR) systems now incorporate 3D viewing systems. AR systems are connected with revolutionary multimedia categories that support natural human interfaces. The Internet of Things (IoT) is a state-of-the-art technology that is used to fetch useful information through sensors. When AR is combined with IoT, especially within a simulation, cost and time in the training process can be minimized. This chapter evaluates the application of sensors in fetching useful information about flights and using that information within flight simulation. Trainees can understand and learn from the simulator just like they would in real flights using AR and IoT. In addition to mitigating simulation, feedback-driven AR systems with sensors can be used for precision data sampling corresponding to flight simulation. Various protocols work for this, such as LEACH and DEEC. The DEEC protocol can be used within flight simulation to achieve reliable data for real-time predictions and is more energy-efficient than the LEACH protocol. The lifetime of sensors is also discussed in this paper along with mitigation strategies.

2.1 Introduction

Specialized skill areas in human life require expensive resources. These skills can only be acquired through practice and time spent in the work area. This chapter presents the application of Augmented Reality (AR) and Virtual Reality (VR) in flight simulation. Flight simulation is generally focused on providing in-depth knowledge of a specific tool or function within a flight. By practicing through a simulator, learners can use the function or tool much more effectively than before. For the simulation of flights in pilot training, sample data must be provided. This can be accomplished through the application of the Internet of Things (IoT). Real-time data used within the simulator can be used to design effective training programs for pilots, which will help them to have hands-on training in real flights. Effective training programs can replace actual aircraft training, and simulation-based flight training could greatly reduce the time associated with aircraft training [1].

Certain conditions must be followed while preparing flight simulation training programs. The design and working of flight simulation must be close to the real-time flight,

Arzoo, Kiranbir Kaur, Salil Bharany, Department of Computer Engineering & Technology, Guru Nanak Dev University, Amritsar 143001, India, e-mails: arzoorana98@gmail.com, kiran.dcse@gndu.ac.in, salil.bharany@gmail.com

https://doi.org/10.1515/9783110785234-002

which is a critical condition to be satisfied by the flight simulator. This can only be possible if an effective data collection mechanism with IoT is available [2]. Data in IoT-driven simulators is obtained with the help of sensors.

Sensors have limited energy associated with them, so energy conservation mechanisms must be in place to ensure their prolonged life. This chapter also sheds light on some of the protocols that can be used to conserve the energy of sensors and prolong the lifetime of the network and better extract data from real-time environments for flight simulators [3].

The Internet of Things (IoT) is one of the greatest technological advances in making electronic devices "smart" by enabling them to operate autonomously. IoT devices are meant to make our lives easier, but they are completely dependent on us. Machine learning (ML) allows user devices to learn usage patterns and preferences. IoT-enabled devices give users access, give commands, and receive instructions from distantly connected devices over the Internet. ML is a machine's learning process that is not explicitly programmed. Machines learn and train from available data. The current limitations of IoT can be overcome with ML. Smart devices no longer require user input to use ML.

There is a large application area associated with smart homes and the Internet of Things (IoT). Client information can be easily extracted, and decisions can be made regarding problems through the use of wearable devices. Sensors attached to the wearable devices can be used to extract information, which is then transmitted through the network.

Figure 2.1 represents a sensor generally placed on wrist or leg bands. These mechanisms are inexpensive and effective for extracting information from the user's body. The data collection mechanism is elaborated in Table 2.1. Data collection through sensors for extracting real-time data can be an effective mechanism for generating user-related data for simulative analysis. Data can be extracted and stored within the cloud with the network as the medium of transfer. Sensor-based mechanisms are implemented within IoT and are commonly used within wireless sensor networks. Nowadays, sensors are also attached to the human body to gather useful information. The gathered information is used for a variety of purposes, such as examining the health condition of the human being based on the gathered information. Information regarding walking, cycling, and other activities can be easily gathered using IoT-based mechanisms. Information regarding the sensors is represented in the following figure.

The data collected from the sensors will be fed into the cloud system. From the cloud, the data will be fetched by the control system and sent to the relevant person for examination. The decision will be made depending upon the machine learning-based mechanism applied. If the result does not fall within a particular class, the activity will be denied.

The chapter is organized as follows. Section 2.1 provides an introduction to flight simulators and their applications in training programs. This section also highlights the need for energy conservation associated with sensors. Section 2.2 describes the applications of AR/VR technology through a literature survey. Section 2.3 discusses the working

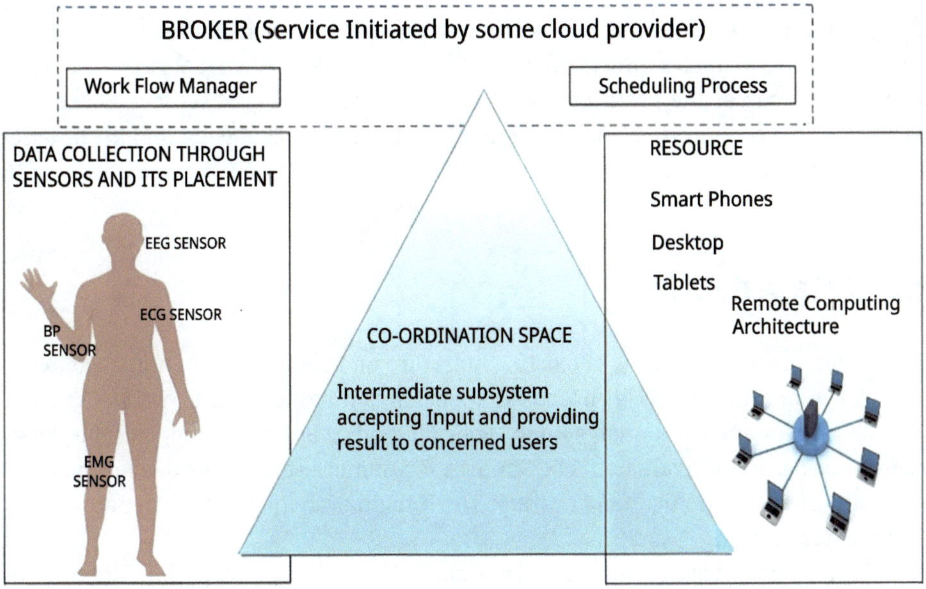

Figure 2.1: IoT usage and sensor placement.

Table 2.1: Parameter Collection settings.

Parameters	**Description**	**Utilization Example**
Human Body	Sensors attached with the human body can generate useful information	Commonly used for disease detection
Home-based environment	Generally attached with the energy-saving devices	Minimizing the energy consumption
Business stores	Transaction management is accomplished through sensors	Abnormality detection within stores and banks
Offices	Used to enhance interaction among intellectuals	Generally used to minimize the energy consumption
Organization like factories, industries, etc.	Used to produce finished goods	Repetitive tasks can be handled using these sensors
Sites where actual work is done	Implemented within the specific customer-based environment	Mining-based applications are generally supported through these sensors
Cars and other moving vehicles	Systems that work inside moving vehicles	Monitoring fuel consumption in vehicles like cars and jeeps
Urban environment	Cities	In Smart Cities for environment monitoring
Miscellaneous	Between urban and rural areas	Including rail tracks, roads etc. used to detect blockage, if any.

of flight simulators, while Section 2.4 provides an analysis of virtual reality-based systems. Section 2.5 provides information about the components of virtual reality systems. Section 2.6 discusses the protocols used to conserve the energy of sensors. Section 2.7 elaborates on a comparative analysis of flight simulation mechanisms, and the last section gives the conclusion.

2.2 Related work

This section presents a review of existing literature on the detection of human activity.

C. Dewi and R. C. Chen [4] proposed a mechanism based on LDA, SVM, KNN, and Random forest to detect human activity. Four methods were compared, and the classification accuracy was evaluated. The Random forest mechanism gave the best possible result in terms of classification accuracy. The human activity recognition dataset was used for comparison.

J. M. Tien [5] discussed real-time decision making with the help of the Internet of Things and artificial intelligence. The decision making could be regarding the user's movement and health monitoring system. The result of the decision is classified according to the confusion matrix.

R. F. Molanes et al. [6] discussed the application of the Internet of Things within deep neural networks. The dataset formed by IoT could be large in nature, and to tackle the issue, layers of the deep neural network must be programmed. This is achieved using this literature.

W. Abdul et al. [7] proposed a visual encryption mechanism using the application of fog computing. The IoT mechanism is enforced within this system to check for the security automatically. The key size and reliability are checked for the security enforcement.

L. H. V. Nakamura [8] proposed an optimization mechanism within IoT. The optimization algorithms including genetic, PSO, and Bees algorithm are demonstrated within this literature. The mechanism shows that the convergence rate is low for these algorithms.

P. Pu et al. [9] discussed how to enable cloud computing within mobile internet of things. The demonstration requires programming sensor-specific applications. Energy conservation is the major concern, and it could be achieved using the cluster-based mechanism.

J. Ploennigs et al. [10] discussed the role of IoT in cognitive buildings. The IoT makes it effective to transfer data from one place to another using the applications of IoT. In addition, the movement of users can be detected by transferring the information using sensors placed within wearable devices.

W. Li et al. [11] proposed a self-learning model for smart homes. The self-learning model allows for easy and fast decision making. Activity monitoring is generally specified using the application of this model.

J. Mohammed et al. [12] proposed patient movement monitoring using the application of webcams and cloud computing. The structured model of cloud computing is used in this model. The mechanism of clustering to send and receive information is used in this case.

H. Velampalli [13] described various techniques for detecting human activity using supervised learning. The techniques described are decision trees, artificial neural networks, multinomial logistic regression, and K-nearest neighbor methods for performing classifications. This improves the classification accuracy of the ANN algorithm. However, the Naive Bayes algorithm is not efficient.

B. Barati and J. Boubana [14] proposed a deep learning approach that helps detect human activity without a time lag. They used sensors to observe human activity in AR/VR and implemented a deep learning mechanism to detect human activity. They classified human behavior using time domain characteristics such as mean, minimum, maximum, variance, and range. The machine learning approach provides higher accuracy. However, it does not work with large datasets.

Z. Gulzar et al. [15] compared different techniques such as ANN, neural networks, and random forests to detect human activity. They used the orange tool to compare the features extracted using different techniques. The classification accuracy is good, but neural networks are not efficient.

Xu et al. [16] proposed a convolutional neuron network-based motion detection mechanism. Support vector machine is used as a classifier. The classification accuracy associated with this approach is relatively high. Large datasets, however, cannot be tackled through this approach.

Sun et al. [17] proposed a low short term memory for the detection of human activity. The benchmark dataset from Kaggle website is used to examine the human activities with AR/VR by performing training and testing processes. The only issue is the handling of larger datasets, as the classification accuracy will be low if larger datasets are used.

Chen et al. [18] proposed a convolution neural network-based mechanism for the detection of human activities with AR/VR. The UCI dataset is used for the detection. The classification accuracy using the neural network-based approach is very high, achieving almost 90 %.

A. Khelalef et al. [19] proposed a deep CNN approach that calculates space-time information of the video and extracts features from it. BSTM (binary space-time maps) is used for operations. The technique is efficient and has a high recognition rate. It is implemented quickly, and computational time is low.

N. Oukrich et al. [20] proposed a supervised learning model for detecting human activities with AR/VR. A minimum redundancy maximum relevance model is used to extract the features and predict human activities with AR/VR. The main problem with this approach is that it depends on the calculations of correlation, and any deviation can affect the classification accuracy.

S. Ha et al. [21] proposed a convolution neural network-based approach for the detection of human activity. Cycling, running, moving, swimming, etc., activities can be easily predicted through this approach. The classification accuracy of this approach is almost 90 %. The preprocessing-based mechanism is missing in this case; hence larger datasets cannot be handled through this approach.

Table 2.2 provides a comparison of different techniques used in existing literature for smart home human activity recognition. It presents the author, technique used, merit, and demerit of each approach.

Table 2.2: Comparison of existing literature of smart homes.

Author	Technique	Merit	Demerit
Y. Chen and Y. Xue [18]	CNN based and utilize a raw tri-axial accelerometer	Computational cost is low and the accuracy rate is high.	The recognition rate is low in a proposed method that must be enhanced.
S. Ha and S. Choi [21]	Supervised learning	Noisy data can be handled easily since relevant data is retained but irrelevant data is eliminated.	The main problem with this approach is that it is depending upon the calculations of correlation and any deviation can affect the classification accuracy.
H. Vellampalli et al. [13]	Supervised learning	Better classification accuracy	Computational cost is high
L. B. Marinho et al. [22]	PDA sensor based human activity recognition	Increased rate of detection	This technique does not involve acknowledgment in the movement, and the data must be collected from various clients.
A. Bevilacqua et al. [23]	CNN model	The combination of various sensors leads to better results and high accuracy.	It is not implemented in the real world.
Sun et al. [17]	LSTM and Elm	Gesture recognition is better; it improves the capability of activity recognition.	Classification accuracy can be further enhanced.
N. Oukrich et al. [20]	BSTM based CNN approach	The technique is efficient and has a high recognition rate. Its computational time is low.	Edge detection is not present.
W. Li et al. [16]	CNN-PFF technique	Classification accuracy of this model is high and performance is better than compared to CNN with ID's kernel.	No edge detection mechanism is presented.
A. Khelalef et al. [19]	CNN based Pruning and Corner detection algorithm	High accuracy and robustness.	It does not utilize feature extraction.
Z. Gulzar et al. [24]	KNN, neural network, random forest	Better recognition rate	The KNN approach is not efficient.
B. Bharathi et al. [25]	Deep learning and machine learning	The machine learning approaches give better accuracy.	It does not work on large datasets.

2.3 Working principles of flight simulators

The flight simulator creates an environment of aircraft where it is implemented. The flight simulator is quite useful for a person who wished to learn how to fly [26]. The person can be a pilot or a normal person. A definition can be taken up as a professional course or just for the sake of entertainment. Flight simulation generally uses prospective projections. The prospective projections indicate that the object will be projected along the lines which will meet at a point. This point is known as a principal reference point. These projections give a realistic view of the object. To support prospective projections the flight simulator must be designed as close as possible to the real-world aircraft. Within the simulator, there exist sensors. These senses detect the movement of the joystick odd steering wheel present within the simulator [27].

As the simulator is rotated sensors will detect that motion and move the object accordingly on the screen. The real-time data with the Internet of Things will continue to enter the simulator. To store the data direct view storage tubes also termed DVST have been used to project real-time situations on the screen, two-electron guns are used these electron guns are known as primary guns and flood guns. The primary gun is used to draw the image on the screen and the flood gun is used to retain the image over the screen. Since DVST stores the image directly within the monitor. Hence the drawing of the image is much faster in this case as compared to CRT-based screens.

2.4 Virtual reality

The virtual reality-based system uses software to create an environment that is similar to the actual system, including visuals and sound. Virtual reality prioritizes the user's visual perception, and real-time data can be presented to generate a realistic working environment [28]. Virtual reality-based systems are now commonly used in gaming platforms, providing a more immersive experience for gamers. Objects move in the direction the gamer moves through the use of sensors. In the field of training, virtual reality has played a significant role, particularly in flight simulator training programs, where expensive resources can be replaced with virtual reality systems.

These systems provide pilots with a realistic flight experience, allowing them to learn how to fly within a limited time frame. Virtual reality-based training programs ensure better training outcomes in less time, which would otherwise require several hours or even days. Data gloves are an excellent example of virtual reality-based systems that record hand motion and move the object on the screen accordingly. Virtual reality-based systems are now also collaborating with machine learning. Using machine learning, virtual reality-based systems can model air traffic control [29], predicting future air traffic and diverting flights to low-congestion paths.

2.5 Components of VR systems

When developing a VR-based system for flights, both hardware and software requirements need to be considered [30]. For a mobile flight simulator, the following components are necessary:

- A nobile phone with android Marshmallow operating system
  This operating system is required to install and use the AR/VR system. Other operating systems can also support AR/VR systems.
- Gear VR
  This is a critical component that the client must have to enter the simulative environment. Sensors attached to the Gear detect the user's motion, and objects move accordingly on the screen. This is made visible with the help of lenses within the Gear control.
- Gamepad
  This controller contains joysticks and buttons, providing multiple options for operating the simulative environment. It is commonly employed in gaming consoles.
- Cloud and IoT
  Both cloud and IoT environments are necessary to fetch and store data from the real environment. This data will be fed into the VR system for processing, and the results will be generated based on the processing within VR. The processing can be based on machine learning, and the complexity can be varied using training modules. Data about the simulation can be stored within the cloud for monitoring.

2.6 Protocols used for conserving the energy of sensors

The information about the flights could be gathered through the sensors. Sensors consume energy as the information is transmitted. The sensor energy conservation is critical otherwise accuracy and reliability of information transmitted could be in question. To this end, the most common protocol used is low energy adaptive clustering hierarchy (LEACH) [31]. The transmitted packets from nodes will be aggregated at the cluster head having the highest energy within this protocol. It is the responsibility of the cluster head to transmit the packets to the base station. The protocols however have shortcomings as a single cluster head could be overloaded and packet loss could be the result. To overcome the issue, distributed energy-efficient clustering protocol was devised. In this protocol, multiple cluster heads per round going to be selected. This will decrease the load on the single cluster head. Energy efficiency thus increases in DEEC [32]. This protocol can be used within flight simulation to achieve reliable data for real-time predictions. In Table 2.3 various comparative analysis of techniques is given.

Table 2.3: Comparative analysis of techniques used in flight simulators.

Author/s	Technique	Application	Merit	Demerit
Y. Li, K. Sun, and X. Li [26]	Principal component analysis	The Architecture of flight simulators is described effectively	Using PCA, critical components required for the simulation of flights were identified	Real-time environment in flight simulation is not considered
L. Zhang et al. [29]	PC-based simulator	PC-based high quality simulator	Low cost and high quality simulation was achieved	No real-time data considered for evaluation
J. W. Wallace et al. [1]	Augmented reality-based flight simulation	Augmented reality-based immersive and tactile flight simulation	War data uploaded within cloud and effectively analyzed through this mechanism	Real time data is not considered
C. Leuze and M. Leuze [2]	Microsoft flight simulator	Microsoft flight simulator for gaming application	Fast and effective gaming applications for entertainment	Other then gaming, it is not demonstrated for other applications
C. Villaciś et al. [28]	Real-time flight simulator for pilot training	Pilots can learn to fly using this simulator	Cheap and effective way to make the pilots earn real-time flight experience	Real-time data from flights and complex situations are not considered

2.7 Comparative analysis

This section presents a comparative analysis of different techniques used in flight simulator applications, along with their merits and demerits.

2.8 Conclusion

In conclusion, AR/VR technology has revolutionized the way we learn and experience different environments. Flight simulators provide an effective way to learn how to fly without using actual aircraft, while human activity monitoring using AR/VR technology has also gained popularity. Convolutional neural networks are the most commonly used technique for detecting activity in AR/VR systems. Energy conservation protocols such as LEACH and DEEC can be used to optimize sensor energy consumption. Cloud-based solutions can be utilized to store progress records and monitor learners' progress. AR/VR technology also has potential applications in driving, traffic simulations, and other fields. Overall, AR/VR-based mechanisms provide cost-effective and accurate ways to learn and experience different applications.

Bibliography

[1] J. W. Wallace, Z. Hu, and D. A. Carroll, "Augmented Reality for Immersive and Tactile Flight Simulation," IEEE Aerospace and Electronic Systems Magazine, vol. 35, no. 12, pp. 6–14, Dec. 2020, https://doi.org/10.1109/MAES.2020.3002000.

[2] C. Leuze and M. Leuze, "Shared Augmented Reality Experience Between a Microsoft Flight Simulator User and a User in the Real World," In Proceedings—2021 IEEE Conference on Virtual Reality and 3D User Interfaces Abstracts and Workshops, VRW 2021, pp. 757–758, Mar. 2021, https://doi.org/10.1109/VRW52623.2021.00261.

[3] C. G. Oh, "Pros and Cons of A VR-based Flight Training Simulator; Empirical Evaluations by Student and Instructor Pilots," In Proceedings of the Human Factors and Ergonomics Society Annual Meeting, vol. 64, no. 1, pp. 193–197, Dec. 2020, https://doi.org/10.1177/1071181320641047.

[4] C. Dewi and R. C. Chen, "Human Activity Recognition Based on Evolution of Features Selection and Random Forest," In Conference Proceedings—IEEE International Conference on Systems, Man and Cybernetics, vol. 2019–Octob, pp. 2496–2501. 2019, https://doi.org/10.1109/SMC.2019.8913868.

[5] J. M. Tien, "Internet of Things, Real-Time Decision Making, and Artificial Intelligence," Annals of Data Science, 2017, https://doi.org/10.1007/s40745-017-0112-5.

[6] R. F. Molanes, K. Amarasinghe, and J. J. Rodriguez-andina, "Deep Learning and Reconfigurable Platforms in the Internet of Things: Challenges and Opportunities in Algorithms and Hardware," IEEE Industrial Electronics Magazine, vol. 12, no. June, pp. 36–49, 2018, https://doi.org/10.1109/MIE.2018.2824843.

[7] W. Abdul, Z. Ali, S. Ghouzali, M. S. Hossain, and S. Member, "Biometric Security Through Visual Encryption for Fog Edge Computing," IEEE Access, vol. 5, 2017.

[8] L. H. V. Nakamura et al., "An Analysis of Optimization Algorithms designed to fully comply with SLA in Cloud Computing," IEEE Latin America Transactions, vol. 15, no. 8, pp. 1497–1505, 2017, https://doi.org/10.1109/TLA.2017.7994798.

[9] P. Pu, "Enabling Cloud-connectivity for Mobile Internet of Things Applications," IEEE Access, 2013, https://doi.org/10.1109/SOSE.2013.33.

[10] J. Ploennigs, S. Member, A. Ba, and M. Barry, "Materializing the Promises of Cognitive IoT: How Cognitive Buildings are Shaping the Way," IEEE, vol. 4662, no. c, pp. 1–8, 2017, https://doi.org/10.1109/JIOT.2017.2755376.

[11] W. Li, T. Logenthiran, S. Member, and V. Phan, "Implemented IoT based Self-learning Home Management System (SHMS) for Singapore," IEEE Access, vol. 4662, no. c, pp. 1–8, 2018, https://doi.org/10.1109/JIOT.2018.2828144.

[12] J. Mohammed, C.-H. Lung, A. Ocneanu, A. Thakral, C. Jones, and A. Adler, "Internet of Things: Remote Patient Monitoring Using Web Services and Cloud Computing," In IEEE, International Conference on Internet of Things (iThings), and IEEE Green Computing and Communications (GreenCom)and IEEE Cyber, Physical and Social Computing (CPSCom), pp. 256–263, Sep. 2014, https://doi.org/10.1109/iThings.2014.45.

[13] H. Vellampalli, "Physical Human Activity Recognition Using Machine Learning Algorithms." 2017.

[14] B. Bharathi and J. Bhuvana, "Human Activity Recognition using Deep and Machine Learning Algorithms," International Journal of Innovative Technology and Exploring Engineering, vol. 9, no. 4, pp. 2460–2466, 2020, https://doi.org/10.35940/ijitee.c8835.029420.

[15] Z. Gulzar, A. A. Leema, and I. Malaserene, "Human Activity Analysis using Machine Learning Classification Techniques," International Journal of Innovative Technology and Exploring Engineering, vol. 9, no. 2, pp. 3252–3258, 2019, https://doi.org/10.35940/ijitee.b7381.129219.

[16] W. Xu, Y. Pang, and Y. Yang, "Human Activity Recognition Based On Convolutional Neural Network," In 2018 24th International Conference on Pattern Recognition (ICPR), pp. 165–170, 2018.

[17] J. Sun, Y. Fu, S. Li, J. He, C. Xu, and L. Tan, "Sequential Human Activity Recognition Based on Deep Convolutional Network and Extreme Learning Machine Using Wearable Sensors," vol. 2018, no. 1, 2018.
[18] Y. Chen and Y. Xue, "A Deep Learning Approach to Human Activity Recognition Based on Single Accelerometer," 2015, https://doi.org/10.1109/SMC.2015.263.
[19] A. Khelalef, F. Ababsa, and N. Benoudjit, "An Efficient Human Activity Recognition Technique Based on Deep Learning," Pattern Recognition and Image Analysis, vol. 29, no. 4, pp. 702–715, 2019, https://doi.org/10.1134/S1054661819040084.
[20] N. Oukrich, E. B. Cherraqi, and A. Maach, "Human Daily Activity Recognition Using Neural Networks and Ontology-Based Activity Representation," In Innovations in Smart Cities and Applications, Lecture Notes in Networks and Systems, vol. 37, pp. 622–633, 2018, https://doi.org/10.1007/978-3-319-74500-8_57.
[21] S. Ha and S. Choi, "Convolutional Neural Networks for Human Activity Recognition using Multiple Accelerometer and Gyroscope Sensors," pp. 381–388, 2016.
[22] L. B. Marinho, A. H. de Souza Junior, and P. P. R. Filho, "A New Approach to Human Activity Recognition Using Machine Learning Techniques," In Advances in Intelligent Systems and Computing, vol. 557, no. December, pp. 529–538, 2017, https://doi.org/10.1007/978-3-319-53480-0_52.
[23] A. Bevilacqua, K. Macdonald, A. Rangarej, V. Widjaya, B. Caulfield, and T. Kechadi, "Human Activity Recognition with Convolutional Neural Networks," 2019, https://doi.org/10.1007/978-3-030-10997-4_33.
[24] Z. Gulzar, A. A. Leema, and I. Malaserene, "Human Activity Analysis using Machine Learning Classification Techniques," International Journal of Innovative Technology and Exploring Engineering, vol. 9, no. 2, pp. 3252–3258, 2019, https://doi.org/10.35940/ijitee.b7381.129219.
[25] B. Bharathi and J. Bhuvana, "Human Activity Recognition using Deep and Machine Learning Algorithms," International Journal of Innovative Technology and Exploring Engineering, vol. 9, no. 4, pp. 2460–2466, 2020, https://doi.org/10.35940/ijitee.c8835.029420.
[26] Y. Li, K. Sun, and X. Li, "General Architecture Design of Flight Simulator Based on HLA," In Proceedings of 2019 IEEE 3rd Information Technology, Networking, Electronic and Automation Control Conference, ITNEC 2019, pp. 2002–2006, Mar. 2019, https://doi.org/10.1109/ITNEC.2019.8728997.
[27] Z. Guozhu and H. Zhehao, "Flight Simulator Architecture and Computer System Design and Research," In 2020 IEEE 2nd International Conference on Circuits and Systems, ICCS 2020, pp. 35–39, Dec. 2020, https://doi.org/10.1109/ICCS51219.2020.9336527.
[28] C. Villaciś et al., "Real-Time Flight Simulator Construction With a Network for Training Pilots Using Mechatronics and Cyber-Physical System Approaches," In IEEE International Conference on Power, Control, Signals and Instrumentation Engineering, ICPCSI 2017, pp. 238–247, Jun. 2018, https://doi.org/10.1109/ICPCSI.2017.8392169.
[29] L. Zhang, H. Jiang, and H. Li, "PC Based High Quality and low Cost Flight Simulator," In Proceedings of the IEEE International Conference on Automation and Logistics, ICAL 2007, pp. 1017–1022, 2007, https://doi.org/10.1109/ICAL.2007.4338716.
[30] A. Schreiber and M. Bruggemann, "Interactive Visualization of Software Components with Virtual Reality Headsets," In Proceedings – 2017 IEEE Working Conference on Software Visualization, VISSOFT 2017, vol. 2017–October, pp. 119–123, Oct. 2017, https://doi.org/10.1109/VISSOFT.2017.20.
[31] P. Li, W. Jiang, H. Xu, and W. Liu, "Energy Optimization Algorithm of Wireless Sensor Networks based on LEACH-B," IEEE Access, 2017, https://doi.org/10.1007/978-3-319-49109-7.
[32] R. Kumar, "Evaluating the Performance of DEEC Variants," IEEE Access, vol. 97, no. 7, pp. 9–16, 2014.

Arihant Singh Verma, Aditya Singh Verma, Sourabh Singh Verma, and Harish Sharma

3 A comprehensive study for recent trends of AR/VR technology in real world scenarios

Abstract: In recent years, various technology giants such as Google, Snap, and Meta (formerly Facebook) have announced plans to invest resources in Virtual Reality products such as the Metaverse. In the growing field of Human-Computer Interaction, the use of Augmented Reality (AR) and Virtual Reality (VR) for interfacing with the digital world is the natural next step. The pandemic prompted a shift to remote work, thereby reducing commute hours and increasing productivity. This shift has provided an impetus for AR/VR solutions to help overcome the difficulties associated with remote work. There has been an uptick in the use of AR in corporate settings, such as conducting meetings, site inspections, and training employees on complex jobs. Mark Zuckerberg mentioned in an interview that Meta's upper management must use this technology for all their meetings. Companies such as SightCall have reported they have over "300 enterprise customers around the world performing more than 3 million customer video assistance calls every year". One of the pioneering trends in AR/VR is its extensive use for gaming. The gaming industry is one of the fastest-growing industries, with 48 % of gaming studios working on AR/VR games. 'Pokémon Go,' a hugely successful AR game, had over 1.1 billion cumulative downloads by 2020, and over 232 million concurrent users at its peak. The rise and development of such games are expected to continue in the coming years, and the VR/AR gaming industry reached a staggering $1.4 billion market size in 2021. Niantic, the creators of 'Pokémon Go,' has made available a Software Development Kit, which can be leveraged by other developers to build their AR products. Technology giant Nvidia offers a development kit, Omniverse, which can be used by other developers to build AR/VR applications. Such initiatives are likely to further bolster development in this field. Our chapter will strategically focus mainly on recent trends and developments in AR/VR technologies, including the Metaverse, AR/VR in gaming, AR/VR developments, and AR/VR in industries.

3.1 Introduction

For years, Extended Reality (XR), which encompasses Augmented Reality (AR), Virtual Reality (VR), and Mixed Reality (MR) technologies, has been on the cusp of widespread

Arihant Singh Verma, Aditya Singh Verma, Sourabh Singh Verma, Harish Sharma, SCIT, Manipal University Jaipur, Jaipur, India, e-mails: arihantsinghverma@gmail.com, asverma314@gmail.com, ssverma80@gmail.com, harish27j@gmail.com

https://doi.org/10.1515/9783110785234-003

adoption. However, it wasn't until the COVID-19 pandemic that XR became an immediate need that fast-tracked development across the industry. Despite its initial adoption being spurred by the pandemic, the benefits of XR will continue even after the pandemic has passed.

Various technology companies such as Google, Snap, and Meta (formerly Facebook) have announced plans to invest resources in Virtual Reality products such as the Metaverse. According to Statista, the global combined market for AR, VR, and MR reached $28 billion in 2021 and is expected to rise to over $250 billion by 2028 [1]. In the growing field of Human-Computer Interaction, the use of AR and VR for interfacing with the digital world is the natural next step.

This chapter is organized into several sections. Section 3.2 focuses on the Metaverse, which is considered the next measure of technology. Section 3.3 discusses the importance of AR/VR, while Sections 3.4, 3.5, 3.6, and 3.7 delve into the current and future possibilities of AR/VR in various fields, such as medicine and gaming industries. Sections 3.8 and 3.9 provide an overview of various tools and gadgets available in the market, followed by Section 3.10, which offers concluding remarks.

3.2 Metaverse could be the Next Big Thing

The original conception of the Metaverse dates back to the early 1990s, in the science-fiction novel Snow Crash [2]. It was imagined as an immersive digital space where people could interact as their animated avatars. Over the years, advances in the development of microprocessors, algorithms for computer graphics, and computer networks have brought us closer to implementing this vision.

The past year has seen an exponential increase in investments related to the development of the Metaverse, with several technology corporations announcing their intent to contribute to this niche. In its present-day form, this technology is envisaged as a digital space that we can enter via a digital VR headset and perform a myriad of activities such as socializing, online shopping, gaming, and virtual travel.

Another critical aspect of the Metaverse is the fact that it must be decentralized, with multiple instances of the Metaverse, owned by different companies or individuals, to be interoperable. This has spurred recent innovations in using blockchain for the implementation of a decentralized protocol, with virtual spaces being represented as blocks [3].

In this direction, chipmakers such as Intel are focusing their efforts on developing new microarchitectures with an emphasis on high-performance computing and visualization to render these rich and immersive graphical experiences [4]. They claim that a distributed and scalable metaverse requires further improvements in computational efficiency, which we are likely to witness in the coming years. They've announced initiatives for the improvement of transistors, VLSI technology, memory, and interconnect at the 2021 IEDM conference [5]. Wearables such as wristbands to track hand motion and gestures form an essential aspect of the ecosystem. Meta Platforms recently announced an AI-powered wristband that gathers input and can adapt itself to the environment [6]. Many other technology organizations have followed suit.

Another emerging area is the development of safety and privacy measures in this digital ecosystem. Various natural language processing algorithms will be integrated to prevent hate speech [7] and bullying [8] in the Metaverse.

3.2.1 Recent investments

Microsoft's recent acquisition of Blizzard, valued at around $68.7 billion, is the largest deal in Microsoft's history. The acquisition was aimed at boosting Microsoft's Metaverse ambitions. Satya Nadella, CEO and chairman, said after the acquisition of the gaming studio, "Gaming is the most dynamic and exciting category in entertainment across all platforms today and will play a key role in the development of Metaverse platforms." [9] Microsoft believes the Metaverse is essentially about creating games, being able to put people, places, and things in a physics engine, and having them relate with others.

Meta, the parent company of Facebook, Instagram, WhatsApp, and Oculus, among other subsidiaries, has also made its intentions clear regarding the recent push into the Metaverse. CEO Mark Zuckerberg says the company now considers itself "Metaverse-first, not Facebook first". Meta also plans to spend $10 billion on developing technology for its Metaverse expansion. In Meta's Founder's Letter, Mark Zuckerberg emphasized that they want to sell their devices at cost or subsidized rates and aim that "within the next decade, the metaverse will reach a billion people, host hundreds of billions of dollars of digital commerce, and support jobs for millions of creators and developers" [10].

Google has also followed suit and invested heavily in the Metaverse, both in private equity and moving its popular platforms towards the Metaverse. AR and popular applications like Maps and YouTube could create a powerful combination in the Metaverse.

Nvidia is another major player investing in the Metaverse. As one of the world's leading GPU developers, most companies would struggle to put out an AR/VR-related product without Nvidia. Therefore, the company is going to be a key part of Metaverse's infrastructure. Nvidia has also launched its Metaverse platform called the Omniverse, which provides real-time 3D design collaboration to millions of creators, designers, and artists to create 3D assets and scenes from the comfort of their laptops or workstation.

3.3 AR/VR in the workplace is a reality

The pandemic prompted a shift to remote work, reducing commute hours and increasing productivity. This shift has provided an impetus for AR/VR solutions to help overcome the difficulties associated with remote work. There has been an uptick in the use of AR/VR in corporate settings, such as conducting meetings, site inspections, and training employees on complex jobs. Mark Zuckerberg mentioned in an interview that Meta's upper management must use this technology for all their meetings.

According to Gartner [11], 40 % of small-to-medium companies are now considering AR/VR, with projections that up to 70 % could do so by the end of 2022. AR/VR is no longer a cutting-edge technology in search of a mainstream application. As employees have become accustomed to video conferencing, interacting in a more immersive XR environment no longer seems like a far-fetched idea.

3.3.1 Truly immersive virtual meetings

What's great about AR/VR meetings is that anyone can participate from anywhere in the world, even from the comfort of their own homes. Even today, superimposing a background on yourself during a Zoom meeting is Augmented Reality. With XR, an individual could present to a meeting or a conference as a 3D hologram instead of a voice on a phone or a face on a screen, write on a virtual whiteboard, make eye contact with attendees, and more. The possibilities are endless.

VR headsets may become the next addition to the workplace. Rather than flying employees out and arranging accommodations for events and meetings, businesses could invest in a VR headset, which typically costs $300 to $500. This would allow everyone to have an immersive experience and interact as if they were there in person. Instead of a select few, the entire workforce could attend and participate in company events.

3.3.2 Smarter, faster training

Training employees using AR/VR is one of the most common use cases prevalent today in the industry. When compared to conventional training, AR/VR-based immersive training can cut training time in half while also improving employee performance by 70 % [12]. By leveraging XR's experiential learning, a factory that began using AR instead of printed instructions saw a 90 % increase in the number of trainees with little or no experience who were able to complete a complex 50-step operation accurately the first time [13].

Similarly, research by Boeing found that AR increased wiring harness assembly productivity by 25 %. Workers at GE Healthcare who were given the task of getting a new picklist order through AR completed it 46 percent faster than when using the traditional method, which depended on a documented list and item searches on a workstation [14]. Case studies from a variety of other industries reveal a 32 percent increase in production on average [15].

The U.S. Bureau of Labor Statistics has already shared concerns that job opportunities in the United States are rapidly surpassing the supply of skilled individuals, leading

Figure 3.1: A person using a VR headset for training (image source: https://unsplash.com/photos/ipDhOQ5gtEk).

to a labor shortage. There is also fear that machines will eventually replace human labor, and that is true in some cases. However, experience at companies that have included AR/VR in the workplace is that, for most jobs, a combination of humans and machines outperforms either working alone as shown in Figure 3.1.

Because of the aforementioned reasons, it is self-evident that AR technologies will play a key role in reducing the skill gap that is causing the scarcity of trained manufacturing workers [16]. AR will enable more workers to do high-skill jobs while also improving their performance, resulting in increased industrial productivity.

3.3.3 Improved remote assistance & effective work environment

AR/VR is increasingly being used for customer support activities. In a traditional field service repair, technicians visit a place and attempt to diagnose the problem based on experience. They enlist the help of an expert if they are unable to resolve the issue on their own. In an AR/VR and AI workflow, technicians use AI machine learning skills to figure out the issue, with the help of other technicians and historical data, in addition to their own ideas. AI verifies the fix and, using an AR overlay, guides the technician through the process.

Companies such as SightCall have reported they have over "300 enterprise customers around the world performing more than 3 million customer video assistance calls every year" [16]. Launched in 2021, SightCall Digital Flows is a self-service tool that allows users to view AR support sessions targeted to solve specific issues without the aid of a customer service professional. L'Oréal, Kraft Heinz, GE Healthcare, and Jaguar Land Rover are among the larger organizations that are now utilizing SightCall Digital Flows to engage with consumers. 'TeamViewer Assist AR', with over a million downloads on the PlayStore, provides on-the-go assistance to identify and troubleshoot real-world problems.

XR solutions provide measurable ROI to businesses in addition to improving consumer experiences. XR is rapidly going from proof-of-concept prototypes to mission-critical jobs as companies reevaluate tactics to create effective and safe work environments for employees. Instead of viewing XR as a one-time solution, organizations will continue to look for ways to use XR and AI to complement or automate human troubleshooting and decision-making to quickly scale and democratize knowledge.

3.4 Evolving mobile AR/VR

Mobile AR/VR can provide immense value to this emerging technology because they are not constrained to a specifically designated environment, unlike modern-day gaming consoles. Additionally, as the user navigates, they can add richness to a variety of novel environments. It makes sense for AR/VR technologies to take advantage of the already-distributed hardware, i. e. mobile phones and tablets. Mobile devices with AR capabilities are becoming cheaper at a rapid pace. There's also a difference between portable devices and mobile devices. Some people are willing to carry additional hardware like VR headsets while others are not.

Some AR mobile implementations are very obvious, like YouTube VR, which provides a fully immersive VR experience using your mobile phone, while other implementations are more “stealth,” in the sense that the user may not even realize that they are engaged in an AR/VR experience, for example, filters on Snapchat.

LiDAR sensors have been around for a long time. They measure distance by timing the light pulse's passage, sending lasers to ping off objects and return to the source. It is a form of time-of-flight camera. These light pulses are not visible to the naked eye. The latest iPhones already have this LiDAR sensor built-in. Using LiDAR, we can scan real-world objects around us.

It will not be long before tiny projectors built into our mobile phones become a reality. Red Hydrogen One had a holographic display built into its 4.7-inch mobile display. Although the implementation of the display was not impressive, products like this push the boundaries of what is possible and set the precedent for more refined implementations in the future. Mobile AR/VR is an area where we could see explosive growth in the coming years.

3.4.1 Computational capacity of mobile devices

A pertinent question that arises here is whether the computational capacity of mobile devices is enough for the advances discussed above to be performant. In this regard, distributed algorithms that can intelligently split the bulk of the processing between the server and the mobile device have proven to be helpful. Zeqi Lai et al., in their paper, demonstrate a split renderer architecture running as a client-server architecture, which is shown to significantly improve performance. They present a framework, Furion [17], which leverages various compression algorithms, parallel processing, and bitrate adaptation to show that the framework can support VR applications on modern-day smartphones without compromising quality and minimizing latency.

3.4.2 Immersive shopping experiences

Even before the COVID-19 pandemic, retail, online courses [18], and online purchasing were in desperate need of a makeover. For instance, consumers did not prefer to buy furniture online because it was often difficult to judge the size without seeing it in person. However, companies are now employing Augmented Reality to allow consumers to virtually project furniture into their homes, allowing them to judge the size and fit. Similar services are now available for eyewear, jewelry, clothing, shoes, and anything else that consumers want to buy.

Nike has taken a different approach by integrating AR and VR into their physical stores. Customers can scan items and enter a VR world where they can experience the supply chain and understand how the item was manufactured.

In collaboration with Facebook, L'Oreal now provides an AR-powered cosmetics try-on experience to its customers. Similarly, during the pandemic, Apple used AR to allow consumers to experience their products from the comfort of their own homes as shown in Figure 3.2.

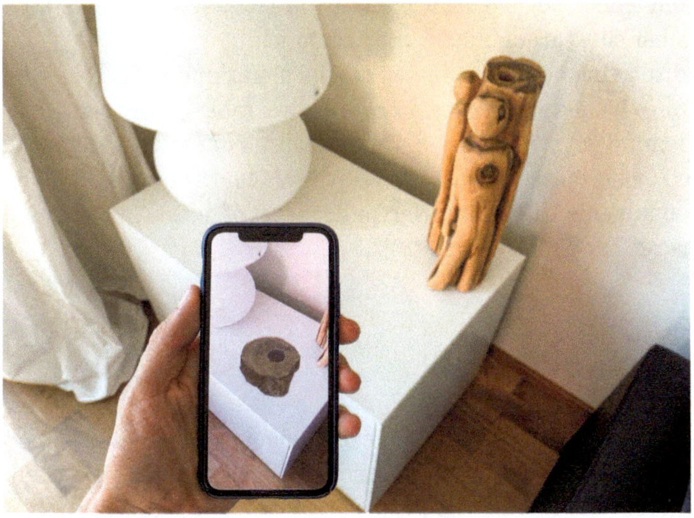

Figure 3.2: Using AR to visualize a product before purchase (image source: https://unsplash.com/photos/NrMGL5MR8uk).

3.4.3 AR in Social Media is here to stay

In social networking, Augmented Reality is becoming increasingly popular. Following Snapchat's lead, Facebook, Instagram, Pinterest, and TikTok have all released their versions of lenses, filters, and effects. AR is proving its worth in the process, not just for marketing and entertainment, but also for generating revenue.

This year, Snap updated its try-on technology to enable more precise eyeglass fittings. It also deployed gesture-recognition technology to allow Snapchat users to signal when they want to see another handbag or see it in a different color. Farfetch is taking advantage of Snap's new voice capabilities to allow users to experiment with different styles while conversing with the app. Consumers' buying activity is also analyzed, so the company can learn which items and styles are popular. Snap's creator program, which paid out $250 million in its first year, is also part of its effort to promote the development of more lenses in the AR space.

3.4.4 Power indoor and outdoor navigation

Augmented Reality has started to have a great deal of impact on Maps and Navigation technology. AR navigation has become more fluid and feasible than ever before. Most crucially, advancements in technology such as Bluetooth Low Energy (BLE) antennas, Wi-Fi RTT, and ultra-wideband (UWB) have made indoor navigation far more feasible than in the past. AR guidance in big indoor settings like distribution centers, retail stores, and airports is one of the most effective implementations of AR Navigation.

Google Maps now lets you use this form of AR Navigation through your phone and navigate complex environments. Google launched this feature for a few cities in March 2021, and expansion to other cities around the world is imminent.

Navigational cues, displaying occluded objects, crucial alerts, and statistics in the field of view of the user on a head-up display of a car, or on wearable glasses can seamlessly integrate into the natural environment [19]. This can bolster efficiency and reduce distractions for the user. Furthermore, AR can aid in travel and tourism as well, by recommending historical monuments, directing toward other areas of interest for the user, and adding context or background information [20].

3.5 Education through AR/VR

Visualization is known to be an effective method for learning new concepts [21]. Augmented and Virtual Reality technologies can become efficacious tools in the field of education and learning, wherein abstract concepts can be easily conveyed to students.

Learning via AR/VR can certainly bridge certain gaps in traditional education; for instance, it enables students to "learn by doing." This is extremely useful in cases where actual work environments might be risky, such as construction, mining, and marine biology. AR/VR is being leveraged with great benefits in the field of online or distance education, where students cannot physically attend lectures.

3.5.1 Fundamental properties

Nick Babich from Adobe details five properties that good AR/VR educational experiences have [22]. These can be used as key points to consider for future developments in this area.

1. **Immersive:** Depending on the context, the XR experience must attempt to simulate reality to the greatest extent possible.
2. **Easy-to-use:** The experience must not be complex and should be inclusive and accessible to individuals and groups of various backgrounds.
3. **Meaningful:** A good XR experience must tell a story and must be relevant to the user.
4. **Adaptable:** Users must be able to control *how* they interact with the app.
5. **Measurable:** The AR/VR tool being used for education must track metrics to capture its efficacy.

3.5.2 Medical education

There has been significant progress in the development of surgery support systems that can assist in operations using overlays that can help guide an operation as well as help medical students master intricate concepts as shown in Figure 3.3.

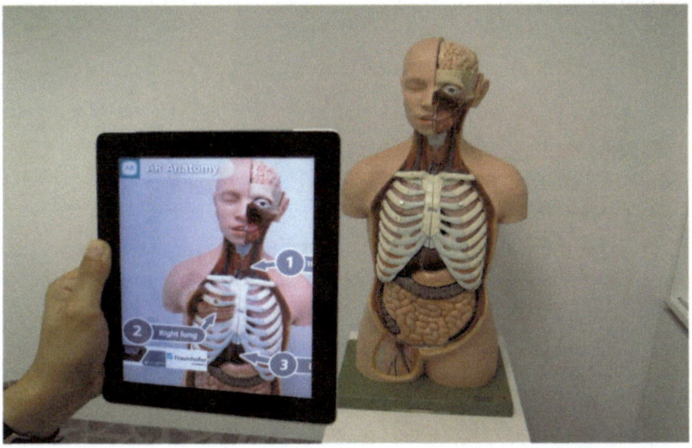

Figure 3.3: Use of AR for medical education (image source: https://pixabay.com/photos/augmented-reality-medical-3d-1957411/).

Currently, medical imaging plays an important role in the diagnosis and prognosis of various diseases [22, 23]. However, certain situations in the medical domain are rare, and training and exposure in a real-life context are not always possible. This is where AR/VR technologies can step in. A paper by Kamphuis et al. provides a thorough analysis of the use of VR in medical education, with the analysis looking quite positive [24].

3.5.3 Research work and academia

A study by Allcoat D. et al. from the University of Warwick demonstrated that students' memorization abilities were measured to be higher when using VR compared to traditional study methodologies. The study participants also reported an increase in positive emotion after the VR learning experience [25]. Another landmark study by Strickland et al. showed how Virtual Reality can help children with autism learn skills in a Virtual Reality setup that they can apply to the real world [26]. An increase in the number of research studies testing AR/VR in the context of education and learning shows an increased interest from the scientific and education community in the adoption of AR/VR technologies. Moreover, the positive results gathered from these studies show that the use of Augmented and Virtual Reality in academia and education is beneficial and will only increase over time.

3.6 Better healthcare with AR/VR

AR/VR has one of the most essential uses in medicine, and it is an industry with very high stakes. The global healthcare Virtual Reality market was expected to be worth $336.9 million in 2020 and is forecasted to increase at a CAGR of 30.7 percent, reaching $2.2 billion by 2024 [https://www.scnsoft.com/virtual-reality/healthcare] as shown in Figure 3.4.

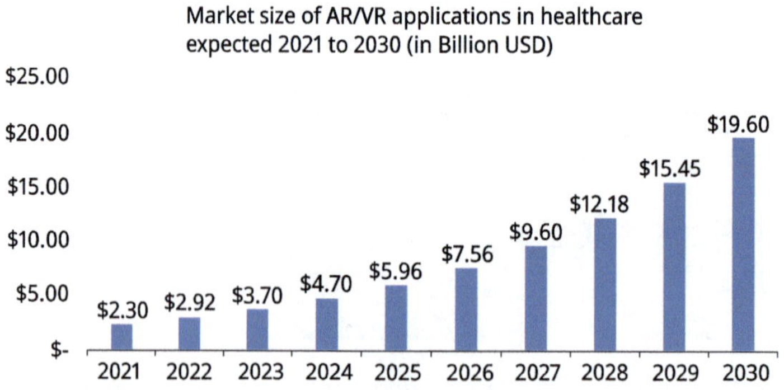

Figure 3.4: Market size of VR applications in healthcare [27].

XR enables you to observe and interact with areas within the human body that would otherwise be inaccessible, thereby bridging the gap between healthcare professionals and patients. Neurosurgeons at Florida Hospital Tampa are using Virtual Reality simulations to visualize the anatomy of a patient's brain. This approach helps patients understand their situation better and make more informed decisions, while doctors can build detailed surgical plans and share them with other clinicians.

Due to the distance to appropriate medical facilities, a patient may only be able to see a non-specialist for a condition. In an immersive experience, XR can bring a specialist physician to the patient. VR is also being employed to view every minute detail of the body and train medical students by creating scenarios that replicate common surgical operations.

Not only is XR reducing the distance to medical facilities, but it is also reducing the distance to new insights. Surgeons are using 3D mapping and images as a "GPS system" to better navigate complicated anatomy and perform more precise surgical treatments. Recently, this technique was used to perform a minimally invasive sinus operation. The system records the operation and surgical planning, which could be used to train other surgeons [28].

In healthcare and medical science, the stakes are significantly higher, and the consequences are much more severe. There is no justification for producing a simplified model of the human body without certain details for performance concerns. The AR/VR tool must be as precise as possible, which is a significant and ultimately productive challenge for AR/VR software developers. This has the potential to change the way the world approaches AR/VR development, making it more responsible and responsive. It is also worth noting that each VR solution for healthcare adds to the study of the long-term impacts of VR use.

3.7 AR/VR is revolutionizing gaming

One of the pioneering trends in AR/VR is its extensive use for gaming. The gaming industry is one of the fastest-growing industries, with 48 % of the gaming studios working on AR/VR games [29]. 'Pokémon Go', a hugely successful AR game, had over 1.1 billion cumulative downloads by 2020, and over 232 million concurrent users at its peak [30]. The rise and development of such games are expected to increase in the coming years. VR/AR gaming industry reached a staggering $1.4 billion market size in 2021 [31] and is expected to expand at a CAGR of 31.4 percent to $53.44 billion in 2028 [32].

According to a quantitative survey done by Ericsson Consumerlabs (Ericsson ConsumerLab Insight Report, 2019), two-thirds of respondents (66 percent) are interested in AR gaming. However, one-third of AR gamers consider that holding a mobile device is insufficient for AR gaming. Additionally, 43 percent of consumers are very interested

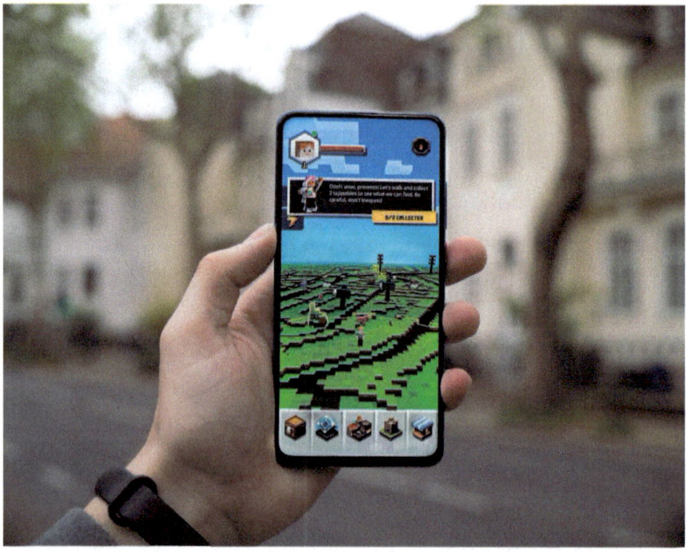

Figure 3.5: A person playing an AR game on their mobile device (image source: https://unsplash.com/photos/sWVAxoLmIzY).

in Augmented Reality sports, and one out of every four customers believes they will use Augmented Reality to exercise in the next five years [33]. Example is shown in Figure 3.5.

AR and VR have improved the gaming experience by allowing gamers to get closer to the action. The sector is predicted to increase due to continuous improvements in existing technology such as motion tracking, 3D effects, and interactive visuals for capturing players' attention. Due to an increase in their buying capacity, users are increasingly demanding new forms of entertainment, leading to a rise in demand for Virtual Reality games.

Resolution Games, a Swedish firm, was one of the first to bet on Virtual Reality gaming. The company's most unique and captivating game to date, Demeo, was released in May 2021. Resolution Games could have easily made Demeo an action role-playing game using Virtual Reality, but instead selected the slower pace of a board game genre that takes us back to analog games like Dungeons & Dragons. One of the reasons for the game's success is that it brought people together to play the game and socialize virtually.

Much of the VR technology that is already permeating other areas such as healthcare and education was developed in the gaming industry. The video game industry is prime for Augmented Reality (AR) and Virtual Reality (VR) applications, with tremendous possibilities to capture brand exposure and consumer loyalty. As seen by hardware sales and software investment, the VR and AR industry has given a completely new dimension to the gaming business, bringing in more than $7.5 billion in revenue by 2020 [34].

3.8 Increasing accessibility and availability of development tools

Community involvement in AR/VR has seen a sharp increase, such as the 250,000 active creators in the Snap AR platform who have developed more than 2 million lenses [35]. As mentioned previously, software development kits such as the Omniverse offered by Nvidia, which can be leveraged to build AR/VR applications, have bolstered development in this field [36]. Niantic, the creators of 'Pokémon Go,' has made available a Software Development Kit to help the community develop AR/VR software. We discuss a few other popular frameworks and toolkits in the following sections.

3.8.1 Vuforia

Introduced by Qualcomm, Vuforia is perhaps one of the most popular AR frameworks. It's an SDK that has seen rapid development and has several features that make it the best for 3D modeling and object recognition. "Ground Plane" helps add content to horizontal surfaces, and "Visual Camera" recognizes and tracks planar graphics and 3D objects in real-time using computer vision technology. Image registration allows developers to position and orient virtual objects, such as 3D models and other media, about real-world objects when viewed through a mobile device's camera.

3.8.2 ARKit by Apple

ARKit was introduced at Apple's Worldwide Developers Conference in 2017. It provides developers with a wide variety of features to create AR applications. The toolkit includes an API for Face Tracking, which enables real-time detection and tracking of up to three faces [37], a feature that has been extensively leveraged by photography and camera apps such as Instagram and Snapchat. Location Anchor is another feature that lets AR creators anchor an AR object to certain geographical coordinates, allowing users visiting the location to experience the digital creation from different perspectives.

ARKit and a few other development kits offer a crucial feature called **Depth API**, which can leverage the LiDAR Scanner to ascertain the per-pixel depth of the surrounding environment. This helps with the placement of virtual objects and opens up a plethora of applications, such as more accurate measurements. Other noteworthy features include the Scene Geometry API, which can help create a 3D map of an environment, and People Occlusion and Motion capture.

3.8.3 ARCore

Google's ARCore is another popular framework for AR development. Compared to Apple's ARKit, ARCore has a larger user base due to the popularity of the Android operating system. According to the official documentation, this framework is based on three fundamental concepts from the field of computational geometry.

3.8.3.1 Motion tracking

ARCore performs sophisticated Simultaneous Localization and Mapping (SLAM), a set of algorithms that maps an unknown environment while simultaneously keeping track of the agent's location. This allows the phone to track its trajectory and alignment relative to the environment [36].

3.8.3.2 Environmental understanding

Alongside motion tracking, the tool implements artificial intelligence techniques that keep detecting feature points and planes in the environment, such as walls, floors, and table surfaces. This information is crucial for object placement in the AR world.

3.8.3.3 Light estimation

The framework can detect average light intensity and gradient in a given frame. This can be leveraged by AR creators who can now illuminate virtual objects using similar light intensity and thus increasing immersiveness.

Apart from these, developers can leverage User Interaction API, Oriented Points, Depth Understanding API, and Anchors and Trackable to develop AR experiences that are immersive and have a greater sense of realism [35].

3.8.4 OpenVR

OpenVR is an open-source software development kit developed by Valve and is compatible with a variety of Virtual Reality headsets. It provides an interface to interact with VR displays without relying on a device-specific development toolkit [37]. Such generalized SDKs are a great resource that developers leverage to develop VR and AR experiences. As of today, the OpenVR repository has more than 5.2k stars and 1.2k forks, a metric that can be used to gauge its increasing popularity and usage.

3.9 Hardware advancements: Making AR/VR more accessible and affordable

Virtual Reality has historically depended on clumsy headsets, high-cost processors, and intricate peripherals to provide realistic experiences. However, recent advancements in underlying hardware are making AR/VR more ubiquitous and accessible.

3.9.1 Cheaper, thinner headsets

Virtual Reality titans can now make headsets that are both cheaper and more powerful than versions from only a couple of years ago. As a result, the demand for headsets is increasing, driving greater innovations and investments in VR hardware. Six million VR headsets were sold in 2021 [38]. Companies are even developing completely new designs to make the next generation of VR technology so lightweight and efficient that it has the potential to entirely change our lives.

Facebook has suggested a new device that regulates the light within a thin lens. This might alleviate the problem of external light interfering with simulated visuals shown to the consumer, lowering the hurdles to lightweight AR/VR integration even more.

'Avegant' has made rapid advances in the miniaturization of AR glasses. By miniaturizing the light engines that produce images behind the lenses of AR glasses, Avegant may have overcome one major obstacle. The AG-50L light engines, which were introduced in September 2021, are as thick as a pencil and weigh roughly the same as a paper clip, according to the manufacturer. The light engines project display light only where the user is looking using eye-tracking sensors. They plan to offer their light engines to

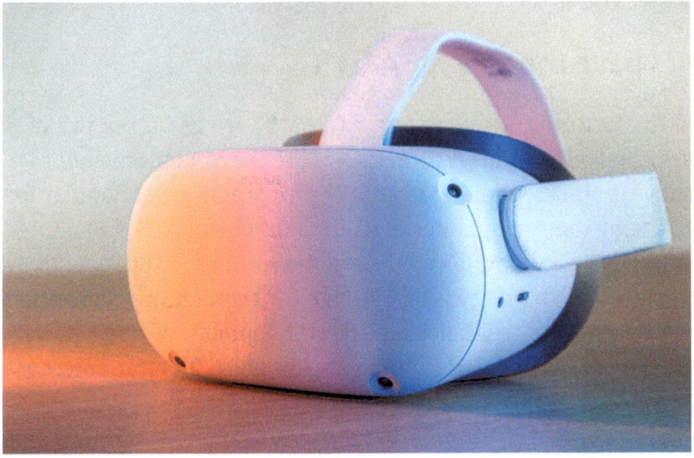

Figure 3.6: A VR headset (image source: https://unsplash.com/photos/Zf0mPf4lG-U).

tech companies who are currently attempting to solve the obstacle of making AR glasses that are sized and weighted similarly to traditional spectacles as shown in Figure 3.6.

HTC has developed an advanced VR headset intended for corporate use, the HTC Vive Focus 3. This unique headgear outperforms consumer VR headsets on the market due to its 5K resolution graphics, accurate head tracking, and 120-degree field of vision. Additionally, HTC has included a fan in the headset, allowing it to drive the CPU hard enough to deliver a more engaging and realistic Augmented Reality experience. However, the HTC Vive Focus 3 is not a consumer product and is geared for business budgets and costs around $1,300.

3.9.2 Improved cloud AR/VR solutions

The transfer of computing power from a local computer to the cloud is the most important aspect of Cloud AR/VR systems. Responsive interactive feedback, real-time cloud-based perception, rendering, and real-time distribution of visual content are all possible with a high-capacity, low-latency internet infrastructure.

Cloud AR/VR is a type of AR/VR that has the ability to change the present business models and expand the cloud platform services currently offered. One example is Nvidia's CloudXR, which allows VR/AR content to be streamed across wireless or wired networks to any end device. With CloudXR, graphics-intensive AR/VR content can be streamed to a low-powered client, such as an Android-based head-mounted display (HMD) or a PC-connected HMD like the Vive Pro, by accessing a high-powered graphics server and streaming over a 5G or Wi-Fi. OpenVR apps may also be streamed to a variety of 5G-connected Android devices, including 5G-enabled phones, using the SDK. This allows for more mobile access to graphics-intensive apps on hardware that is low-powered.

3.9.3 Powerful GPUs pave the way for AR/VR

AR/VR is the future, and the future of AR/VR is faster GPUs! There are three primary players in the GPU market: Intel, AMD, and NVIDIA. Intel is the most popular vendor for integrated, low-performance graphics, while AMD and NVIDIA are the most popular for high-performance GPUs. Among these companies, NVIDIA dominates the AR/VR market.

Today, combining CPU and GPU computing hardware and software is a popular way to get the best of both worlds. This trend towards CPU-like GPU computing is expected to continue in future general-purpose technologies. The development of more powerful processors and GPUs each year is likely to continue making AR/VR better and more accessible in the years to come as depicted in Figure 3.7.

Figure 3.7: Graphical Processing Units (image source: https://pixabay.com/photos/nvidia-graphic-card-rtx-gtx-1080-5264914/).

3.9.4 Surveys and statistics

According to statistics, the demand for headsets in the market steadily increased between Q4 2019 to Q4 2020 [39], and this demand is expected to rise, showing a growing trend in the use of headsets for AR/VR experiences.

In a survey conducted over 2018 and 2019, with a forecast until 2023, it was concluded that spending on AR/VR for onsite assembly and safety purposes is projected to grow with a compound annual growth rate of 177.4 percent, while spending on Augmented Reality games is set to grow at a CAGR of 175.9 percent [40].

3.10 Conclusion

AR/VR technologies are poised to become a primary means of interfacing with the digital world in the future. The development of XR is progressing in all aspects, ranging from hardware, such as improvements in microprocessors, to software. This makes the field a hotbed of innovation.

The statistics, research articles, and market movements outlined in the preceding sections provide a clear view of the recent trends in the fields of Augmented Reality, Virtual Reality, and Mixed Reality. There are clear indications that most, if not all, of the trends discussed in this chapter have gained momentum over the past couple of years and will continue to show positive growth in the upcoming years.

Bibliography

[1] Augmented Reality (AR), Virtual Reality (VR), and Mixed Reality (MR) Market Size Worldwide in 2021 and 2028. Statista. https://www.statista.com/statistics/591181/global-augmented-virtual-reality-market-size/ (accessed April 14, 2022).

[2] K. Orland, "So What Is 'the Metaverse,' Exactly?." Ars Technica, 2021. https://arstechnica.com/gaming/2021/11/everyone-pitching-the-metaverse-has-a-different-idea-of-what-it-is/ (accessed April 14, 2022).

[3] B. Ryskeldiev, Y. Ochiai, M. Cohen, and J. Herder, "Distributed Metaverse. Creating Decentralized Blockchain-based Model for Peer-to-peer Sharing of Virtual Spaces for Mixed Reality Applications," In Proceedings of the 9th Augmented Human International Conference, Article 39, pp. 1–3, Association for Computing Machinery, New York, NY, USA, 2018, https://doi.org/10.1145/3174910.3174952 (accessed April 14, 2022).

[4] "Powering the Metaverse." Intel. https://www.intel.com/content/www/us/en/newsroom/opinion/powering-metaverse.html#gs.wv3qje (accessed April 14, 2022).

[5] "Intel Breakthroughs Propel Moore's Law Beyond 2025." Intel. https://www.intel.com/content/www/us/en/newsroom/news/intel-components-research-looks-beyond-2025.html#gs.wt9jna (accessed April 14, 2022).

[6] "Inside Facebook Reality Labs." The Next Era of Human-Computer Interaction. https://tech.fb.com/ar-vr/2021/03/inside-facebook-reality-labs-the-next-era-of-human-computer-interaction/ (accessed April 14, 2022).

[7] Z. Al-Makhadmeh and A. Tolba, "Automatic Hate Speech Detection Using Killer Natural Language Processing Optimizing Ensemble Deep Learning Approach," Computing, vol. 102, pp. 501–522, 2020, https://doi.org/10.1007/s00607-019-00745-0 (accessed April 14, 2022).

[8] Cyberbullying Detection in Social Networks Using Deep Learning Based Models; A Reproducibility Study – arXiv:1812.08046v1 [cs.CL]. https://doi.org/10.48550/arXiv.1812.08046 (accessed April 14, 2022).

[9] Microsoft to Acquire Activision Blizzard to Bring the Joy and Community of Gaming to Everyone, Across Every Device. Microsoft. https://news.microsoft.com/2022/01/18/microsoft-to-acquire-activision-blizzard-to-bring-the-joy-and-community-of-gaming-to-everyone-across-every-device/ (accessed April 14, 2022).

[10] "Founder's Letter." Meta, 2021. https://about.fb.com/news/2021/10/founders-letter/ (accessed April 14, 2022).

[11] "Augmented Reality in 2020 – It's Time to Get Familiar." Troia. https://www.troia.eu/news/ID/355/Augmented-Reality-in-2020-%E2%80%93-It-is-time-to-get-familiar (accessed April 14, 2022).

[12] S. Rogers, "How VR, AR, And MR Are Making A Positive Impact On Enterprise." Forbes. https://www.forbes.com/sites/solrogers/2019/05/09/how-vr-ar-and-mr-are-making-a-positive-impact-onenterprise/#d463c955253f (accessed April 14, 2022).

[13] M. E. Porter and J. E. Heppelmann, "Why Every Organization Needs an Augmented Reality Strategy." Harvard Business School. https://www.hbs.edu/faculty/Pages/item.aspx?num=53458 (accessed April 14, 2022).

[14] Upskill & GE Healthcare. Pick & Pack Productivity Study Using Skylight. YouTube. https://www.youtube.com/watch?v=AwZ3yYydOH4 (accessed April 14, 2022).

[15] Augmented Reality Is Already Improving Worker Performance. HBR. https://hbr.org/2017/03/augmented-reality-is-already-improving-worker-performance (accessed April 14, 2022).

[16] The 10 Most Innovative Augmented Reality and Virtual Reality Companies of 2022. Fast Company. https://www.fastcompany.com/90715451/most-innovative-companies-augmented-reality-virtual-reality-2022 (accessed April 14, 2022).

[17] Z. Lai, Y. C. Hu, Y. Cui, L. Sun, N. Dai, and H.-S. Lee, "Furion: Engineering High-Quality Immersive Virtual Reality on Today's Mobile Devices," IEEE Transactions on Mobile Computing, vol. 19, no. 7, pp. 1586–1602, 2020. IEEE. https://ieeexplore.ieee.org/document/8700215 (accessed April 14, 2022).

[18] H. Sharma, S. S. Verma, V. Sharma, and A. Prasad, "Impact and Challenges for the Indian Education System Due to the COVID-19 Pandemic," In Impacts and Implications of COVID-19. An Analytical and Empirical Study, pp. 63–85, 2021.
[19] W. Narzt, G. Pomberger, A. Ferscha, et al., "Augmented Reality Navigation Systems," Universal Access in the Information Society, vol. 4, pp. 177–187, 2006, https://doi.org/10.1007/s10209-005-0017-5 (accessed April 14, 2022).
[20] T. Höllerer and S. Feiner, "Mobile Augmented Reality." Telegeoinformatics. Location-based computing and services 21. https://sites.cs.ucsb.edu/~holl/pubs/hollerer-2004-tandf.pdf (accessed April 14, 2022).
[21] J. K. Gilbert, "Visualization. A Metacognitive Skill in Science and Science Education," In Visualization in Science Education. J. K. Gilbert, editor. Models and Modeling in Science Education vol. 1. Springer, Dordrecht, 2005, https://doi.org/10.1007/1-4020-3613-2_2 (accessed April 14, 2022).
[22] S. S. Verma, A. Prasad, and A. Kumar. "CovXmlc: High Performance COVID-19 Detection on X-ray Images Using Multi-Model Classification," Biomedical Signal Processing and Control, 2022. https://doi.org/10.1016/j.bspc.2021.103272.
[23] S. S. Verma, S. K. Vishwakarma, and A. K. Sharma, "A Review on COVID-19 Diagnosis Using Imaging and Artificial Intelligence," In Innovations in Information and Communication Technologies (IICT-2020), pp. 295–300. Springer, Cham, 2021.
[24] How VR in Education Will Change How We Learn and Teach. https://xd.adobe.com/ideas/principles/emerging-technology/virtual-reality-will-change-learn-teach/ (accessed April 14, 2022).
[25] C. Kamphuis, E. Barsom, M. Schijven, et al., "Augmented Reality in Medical Education?" Perspectives on Medical Education, vol. 3, pp. 300–311, 2014, https://doi.org/10.1007/s40037-013-0107-7 (accessed April 14, 2022).
[26] D. Allcoat and A. von Mühlenen. "Learning in Virtual Reality. Effects on Performance, Emotion, and Engagement," Research in Learning Technology, vol. 26, 2018, https://doi.org/10.25304/rlt.v26.2140 (accessed April 14, 2022).
[27] The Global Virtual Reality (VR) in Healthcare Market Is Projected to Grow from $1,206.6 Million in 2021 to $11,657.8 Million in 2028 at a CAGR of 38.3 %. https://www.fortunebusinessinsights.com/industry-reports/virtual-reality-vr-in-healthcare-market-101679 (accessed April 14, 2022).
[28] Texas Surgeons Perform First Sinus Surgery Using AR. https://www.mobihealthnews.com/content/texas-surgeons-perform-first-sinus-surgery-using-ar?_lrsc=9abea7e0-89b4-4d8b-9a81-f2038eb7b910 (accessed April 14, 2022).
[29] GDC. Game Devs Focus on Unionization, Fighting Toxicity, and Adopting Blockchain/Metaverse. https://venturebeat.com/2022/01/20/gdc-game-devs-focus-on-unionization-fighting-toxicity-and-adopting-blockchain-metaverse/ (accessed April 14, 2022).
[30] Pokémon Go Revenue and Usage Statistics, 2022. https://www.businessofapps.com/data/pokemon-go-statistics/ (accessed April 14, 2022).
[31] Virtual Reality (VR) – Statistics & Facts. Statista. https://www.statista.com/topics/2532/virtual-reality-vr/ (accessed April 14, 2022).
[32] AR and VR in Gaming is Projected to Grow up to US $53.44 Billion by 2028. https://www.analyticsinsight.net/ar-and-vr-in-gaming-is-projected-to-grow-up-to-us53-44-billion-by-2028/ (accessed April 14, 2022).
[33] Ericsson ConsumerLab insight report, 2019. https://www.ericsson.com/en/reports-and-papers/consumerlab (accessed April 14, 2022).
[34] What Is the Impact of AR and VR on the Gaming Industry? https://pixelplex.io/blog/ar-and-vr-in-gaming/ (accessed April 14, 2022).
[35] Dream It. Build It. Snapchat. https://ar.snap.com/ (accessed April 14, 2022).
[36] A New Era of 3D Design Collaboration and World Simulation. Nvidia. https://www.nvidia.com/en-us/omniverse/ (accessed April 14, 2022).

[37] More to Explore with ARKit 5. https://developer.apple.com/augmented-reality/arkit/ (accessed April 14, 2022).
[38] Overview of ARCore and Supported Development Environments. https://developers.google.com/ar/develop (accessed April 14, 2022).
[39] Compound Annual Growth Rate (CAGR) of Augmented and Virtual Reality (AR/VR) Use Case Spending Worldwide From 2018 to 2023. https://www.statista.com/statistics/1012881/worldwide-ar-vr-spending-use-case-cagr/ (accessed April 14, 2022).
[40] Augmented Reality Trends. https://lumusvision.com/augmented-reality-trends-infographic/ (accessed April 14, 2022).

Medini Gupta and Sarvesh Tanwar

4 AR/VR boosting up digital twins for smart future in industrial automation

Abstract: In the last few decades, globalization has been a major factor driving the adoption of digital twins for innovative manufacturing methods. Industrial operators work on product design, operations, and post-sales demands, and enabling digital twins in industrial activities opens doors for new opportunities. Collaborating data collected from human beings with smart machines can uncover distinct ways to redesign the backbone of the automation industry. A digital twin is a virtual depiction of physical entities, including people, processes, and devices. This technology has gained much recognition in detecting disruptions, thus enhancing logistics and manufacturing with quick decision making. However, the ease of use for both AR/VR gadgets is still low, with short battery life and limited accuracy, which poses a challenge for wider acceptance. Technological development is indispensable, and cloud services support distributed working schedules for manufacturing. A digital mirror of the real entity can be used in manufacturing to monitor the performance and predict the health of assets. Technical assistants and end-users must work together to derive the full potential of evolving technologies. This paper aims to explore the role of AR/VR with digital twins in strengthening the automation industry, providing a comprehensive view of AR/VR and the workflow of digital twins, addressing the existing obstacles of the automation industry, and focusing on the business use cases of these technologies. We reviewed the literature to gain insight into the factors that lead organizations to implement digital twins with AR and VR in manufacturing, the components that allow these solutions to be executed successfully, and the barriers responsible for slowing down the widespread acceptance of digital twin solutions.

4.1 AR/VR for complete view

Augmented reality magnifies the physical world by overlaying layers of virtual elements onto it, in the form of computer graphics, haptics, audio-visual or sensor projects. The main goal behind adding this virtual information is to deliver an appealing and immersive end-user experience by receiving input from different devices such as smartphones, smart goggles, and smart lenses. One of the most popular uses of augmented reality was seen in 2016 when Pokemon Go became an international sensation [1]. In this game, users have to track and capture Pokemon characters that appear in the physical world.

Medini Gupta, Sarvesh Tanwar, Amity Institute of Information Technology, Amity University Uttar Pradesh, Noida, India, e-mails: guptamedini642@gmail.com, s.tanwar1521@gmail.com

https://doi.org/10.1515/9783110785234-004

Augmented reality highlights the particular characteristics of physical surroundings, provides better details of those characteristics, and shares smart insights in the form of output that can be applied in the natural world [4].

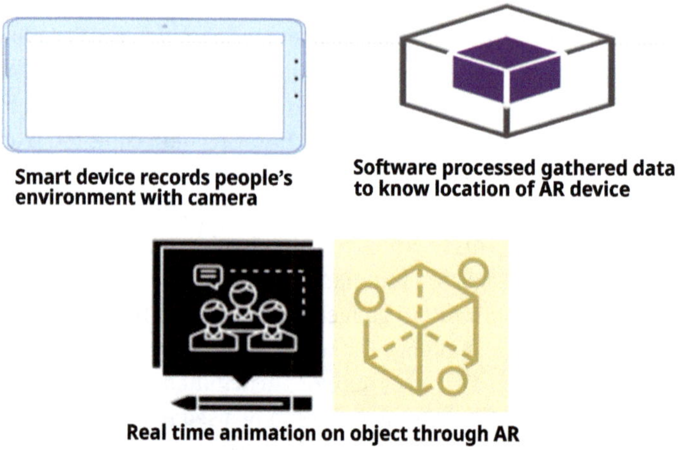

Figure 4.1: Augmented Reality.

Virtual reality is a computer-provided environment that can be the same as the physical surrounding or entirely different, and it is customized in such a way that users are made to believe it to be real. We often come across the application of virtual reality with respect to 3-dimensional images that can be experienced on a computer by operating a mouse or keyboard to change the direction or zoom the image [2]. Immersive head-mounted devices are virtual reality smart display equipment that employs an optical system to present virtual surroundings by taking input from a display produced by a computer in front of the end-user's eyes [22]. The illusion of being physically present is created by motion sensors that capture people's movements and regulate the scene on the display device consistently.

We have implemented digital twins on the Microsoft Azure environment to explore digital twins. Pre-built graphs are uploaded to the platform to gain insights about the digital twin model. Digital twin properties of each model are visualized, and query language is used to get responses from the digital twin environment. Real-time data updated by IoT sensors, as well as manually entered details for twin properties, can be edited.

The rest of the paper consists of the following sections: Section 4.2 discusses the introduction to digital twins along with their workflow. Section 4.3 deals with the obstacles in the manufacturing industry that require immediate attention, and proper solutions can be achieved with technological advancements. Section 4.4 presents the related work undertaken by researchers in this field. Section 4.5 talks about cloud-based digital twins, taking Microsoft Azure as a platform to deploy DT models. The significance of AR, VR, and digital twins in real-world business aspects has been highlighted in Section 4.6. Numerous opportunities that can be derived by converging these three cutting-edge technologies are mentioned in Section 4.7. Practical implementation to visualize digital twins is shown in Section 4.8, where pre-existing graph samples are taken into consideration to deploy the model on the Azure platform. We conclude the paper by stating the promising outcomes that would be gained once digital twins, VR, and AR are merged and executed successfully. However, a large amount of investment is needed by the corporate sector to experience widespread adoption.

Motivation

The widespread pandemic that still continues to disrupt businesses has prompted organizations to search for new ways to minimize their risk for financial loss [3]. During the initial months of the pandemic, organizations struggled to fulfill demand, which further expanded their risk for financial loss. Digital twins, connected with simulation ability, allow organizations to execute an endless number of what-if situations to identify potential challenges and golden opportunities. In this way, stakeholders experience minimal loss when decision capacity gets mishandled. The level of visualization and predictability that digital twins carry along is not found in traditional planning software. It is human nature to aspire to experience more than what we already have [8]. Due to this, numerous innovations and technological developments have taken place, made possible through augmented reality, where surroundings can be designed in a way that we want to see in the physical world. Augmented reality has become the talk of the town after the rise of the metaverse. Elements can be viewed from various angles, providing a better understanding of how that element really looks in a 3-dimensional environment. Virtual reality primarily focuses on sound and vision. The audience experiences an immersive environment in a digital world created by computer simulation while wearing a headphone. Virtual reality has introduced multiple options in media advertisement [5]. Other industrial sectors are also reaping the advantages of virtual reality. The e-commerce sector has experienced large end-user involvement by applying this technology.

4.2 Digital twins

A digital twin is a virtual representation of real-life equipment or a process. This technology is used to replicate processes by collecting data to forecast how the equipment will perform. Simulations for replicating the physical world are built by integrating digital twins with Artificial Intelligence (AI) and the Internet of Things (IoT) [5]. Progress in Machine Learning (ML) algorithms and data analytics techniques have given rise to digital models that provide enhanced performance and innovative solutions. IoT smart sensors are connected to physical objects, and based on the data collected, required modifications are made to the virtual replica. Decision-making based on data has allowed IT professionals to improve the efficiency and lower the cost of machinery [3]. The concept of digital twins has not only covered cities, buildings, jet engines, bridges, or factories but has also expanded its advancements towards people and processes.

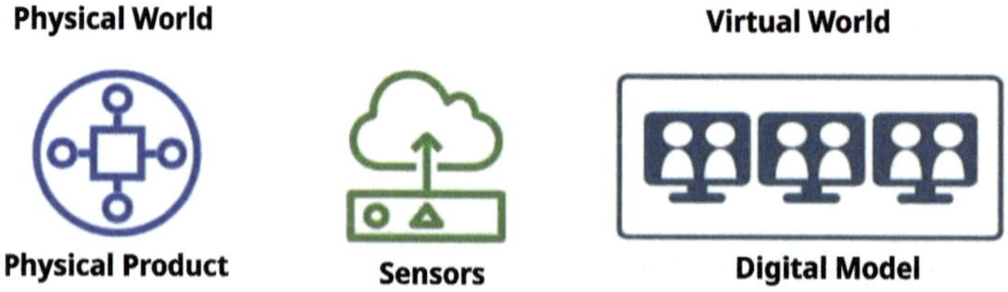

Figure 4.2: Digital twin.

4.2.1 Workflow of digital twins

Visualization algorithms are deployed that allow developers to detect inner challenges of a real product without physical interference, thus not taking any risks regarding the safety and health of employees [21]. Flaws can be reduced during production when developers perform testing and implementation procedures on its digital counterpart [6]. This is a quick and low-cost methodology to identify any flaws in the virtual product as compared to its physical part.

4.2.1.1 Data aggregation

Engineers collect data of various categories about the particular product to work on, such as physical composition, outward appearance, reaction under particular circumstances, etc.

4.2.1.2 Modelling

Using the collected data, computer software is used for modelling data, and a mathematical model is built that correctly depicts the features of the physical product. The mathematical model will be an exact replica of the parallel product – starting from small particulars to the behavior of that product, which will be the same as the physical product [11].

4.2.1.3 Integration

In the last step, integration of the original product with its digital counterpart takes place to enable real-time monitoring. This is achieved by attaching sensors to the real product, and data is transmitted to the respective IoT platform for further visualization [14].

4.3 Obstacles in the manufacturing industry

The manufacturing industry works on processing raw materials into semi-finished or complete furnished goods to make them available for consumers [10]. The rise of the manufacturing industry is also affected by the growth in technology with the use of smart grids. Compulsive manufacturing software permits industries to enlarge their business productivity. Applying technologies in manufacturing with computers and robots have increased its applications during the pandemic [7]. Data-based decision-making on machines will lead towards labor-intensive, time and resource-efficient, and will also bring down financial expenditure at the same time. Here we have discussed what obstacles in the manufacturing sector need to be addressed with technological solutions.

4.3.1 Supply Chain Disruption

Moving raw materials from suppliers to distributors and then delivering the product to consumers. Indecency in any stage of the supply chain will affect the entire production procedure. This will not only cause momentary loss but also put the health of stake-holders at risk. Manufacturers should make sure that goods are bought from reputable providers, and customers should receive their product on time and in proper condition [8].

4.3.2 Product uniformity

End-consumers prefer their products to be bought from reputed brands to get good quality of goods and services. Industries that deal with pharmaceuticals, beverages, chemicals or other goods that involve process manufacturing need to be more careful [12]. Brand reputation continues to rise if consumers get proper service, but at the same time, if any casual deviation occurs, it directly affects safety and brand consistency.

4.3.3 Material waste

Increased efficiency, balancing profit and loss, managing high overhead can be achieved by lowering down the wastage of material. Currently, users are more attracted to eco-friendly goods. Industries involving chemical processes need to be alert about the composition they are utilizing and what will be its ill-effects on the environment [9].

4.4 Related works

There is a growing body of research focused on identifying, reviewing, and proposing solutions for the application, benefits, and challenges of augmented reality (AR) and virtual reality (VR) with digital twins in various domains. In this section, we review several research papers related to the manufacturing industry.

Eleonora Bottani et al. [1] presented a review of emerging applications of AR in various industries, including education, manufacturing, aircraft, and automation. The authors analyzed over 170 papers to identify different fields where AR is being used. They found that tablets, cameras, smartphones, and HMDs are being adopted in many settings to experience augmented reality. Most of the application papers focused on the machine tools industry or the general context of manufacturing.

Wenjin Tao et al. [2] focused on assembly simulation in manufacturing with AR/VR. They reviewed sensing, modeling, and interaction techniques that lead to assembly simulation using AR and VR. The authors also discussed the acquisition of virtual data from a physical asset and the development of a 3D model using collected data. Multiple methods for AR tracking and enabling human-computer interaction with audio and visual platforms were also discussed. The authors provided a case study for practical hands-on experience with AR for assembly simulation.

Åsa Fast-Berglund et al. [3] presented the results of five case studies focused on virtual reality in the maintenance field. The authors found that VR is more prominent in the maintenance phase, as it provides immersive training experiences. However, VR is less useful in operational phases. The authors also noted that organizations are rapidly adopting digitalization, and smart wearables and applications have become a new way of life. They built a digital assembly training device for using gearboxes and used head-worn immersive glasses in two of the five case studies on VR.

Edward Kosasih et al. [4] reviewed the main reasons causing delays in the early implementation of digital twins. The authors found that forecasting and real-time management in digital twins is regulated by big data, machine learning, and IoT. Digital twins can be costly if the life expectancy of the project is short, and interoperability among different elements in real-time can take a long time. The amount of learning required to build a physical asset is the same for the digital counterpart.

Costantini et al. [5] presented their research on developing a hybrid digital twin model that is easily approachable by manufacturers to deploy their services. They used big data based on cloud to design digital twin applications and take advantage of cloud services. For the manufacturing sector to embrace digitalization of their tools and products, it is necessary to utilize a software-based approach. The authors proposed an architecture consisting of data handling, computational devices, anonymization for maintaining confidentiality, and encryption.

Anand et al. [25] discussed the popularity of AR and VR equipment and their potential use cases in the tourism industry. The authors reviewed the success rate and challenges for further development and acceptance of VR and AR applications to enhance marketing. Tracking sensors enable customers to move around in a physical area and update their location in the digital world with the help of AR and VR headsets. AI-derived chatbots have helped the hospitality sector by resolving customer queries, which reduces repetition and time consumption by customer personnel.

Michael Weinmann et al. [26] proposed a unique 3D labeling system focused on VR. The central server monitors the 3D model and handles clients and queries. A large number of end-users are able to mark various scenes in parallel. The system also has great opportunities in education, virtual consultation, and disaster management. Misalignment and artifacts can pose challenges due to uninterpreted camera movements.

Noureddine Elmqaddem [27] discusses the convergence of VR and AR in educational institutes. The development of computing capabilities has shown significant improvement in the successful deployment of these technologies for learning purposes. New innovative methodologies should fulfill the demands of the current era. Giant companies like Microsoft and Google have invested in these technologies to further brighten the future.

4.5 Azure based digital twins

The Microsoft Azure-based Digital Twin (DT) environment is one of the most commonly used platforms to deploy DT solutions. Virtual models of entire surroundings, including bridges, factories, houses, buildings, roadways, and farms, can be created with twin graphs [8]. End users can have customized experiences with reduced costs and enhanced operations as DT provides insights about the product that drive accurate decision-making [10]. Modeling languages are developed to easily build virtual replicas of smart surroundings. Doosan Heavy designed smart wind farms by deploying digital twins, which allow employees to manage the performance of appliances from a remote location. Depending on meteorological conditions, estimates of energy produced by wind farms are made. The update and monitoring of IoT devices can be achieved with much-needed scalability and security using Azure solutions converged with IoT [7].

4.5.1 Real-time visualization of IoT services

An advanced IoT model consists of a very large number of interconnected devices. Managing and debugging these devices in such a large margin becomes a huge task [14]. Upgradation of IoT devices in the future can be identified timely. Digital twins with Azure enable users to sustain a complete view of the system that delivers real-time visualization of the system [15]. The feature that allows users to make use of an already existing hub turns out to be easier to add Azure with digital twins.

4.5.2 Process analysis of data

Integrating Microsoft's cloud with digital twins allows end-users to access live information from Azure architecture to utilize cloud services for further analysis. Connect products such as CRM, ERP, and IoT edge gadgets with digital twins to promote real-time execution [16].

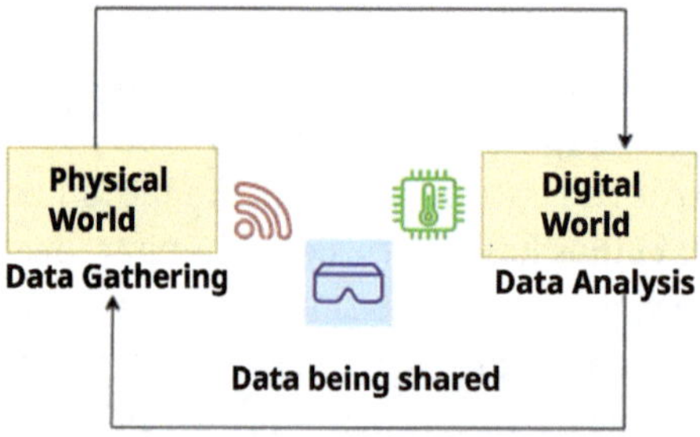

Figure 4.3: Digital twin architecture.

4.6 Business aspects of AR, VR and digital twins

AR-VR provides restorative visuals and adds a enhanced layers in various sectors such as manufacturing, healthcare and education. Complicated surgical operations are being performed by doctors with these technologies. Wearables based on AR-VR in manufacturing sector monitor the changes, detect risky working environment and predict the design structure. Educational institutes have employed augmented reality and virtual reality to make learning more interactive and innovative [15]. Organizations obtains effective insights regarding performance of assets, upgrade quality of assets, enhance consumer experience and bring downs operational price.

4.6.1 Use of AR at the National Water Company of Israel

The National Water Company in Israel has implemented the use of AR glasses and a smartphone application for visualizing marketing and diagrams. Employees of the company are equipped with smart glasses [13]. The application assists and monitors the employees, providing specific guidance and instructions when electricity installation is taking place at the Mekorot facility.

4.6.2 NuEyes

NuEyes are lightweight AR smart glasses that are worn on the head and activated by voice for the visually impaired. The product is easy to control with a wireless remote controller, and a video camera is installed at the front of the glasses that enlarges what the user is looking at and displays it inside the smart glasses. The user can read, write, watch television, and even see human faces [14].

4.6.3 Aquas do Porto

This Portuguese organization manages the water supply of the city of Porto. Digital Twins are used to predict floods, address problems related to water quality, and overall improve water infrastructure. A digital replica is created using telemetry information, sensors, maintenance, asset management, account management, billing, etc. Potential leaks are quickly detected, reducing water waste [18].

4.6.4 Unilever PLC

Unilever PLC has implemented digital twins to provide flexible and effective production procedures. Digital models of the factories are built with digital twins. Sensors share live performance data, including the speed of motors and temperature, on a cloud platform at every geographical location. Machine learning techniques are used for identifying and tackling emergency scenarios [18].

4.6.5 Virtalis

The UK police have implemented a VR-based training program that is mandatory for newly appointed officers [20]. The VR-based training program eliminates the need for physical visits to crime locations by the officers until they are fully trained. Smart headsets are portable and provide remote learning, which is an added advantage during the pandemic. Resources and cost savings are achieved as no physical instructors are required for training [26].

4.6.6 BreakRoom

BreakRoom is a VR workspace where employees can customize a work environment that is comfortable, productive, and maintains focus. It consists of a headset and noise-canceling headphones. Users can turn off the background sound as per their need, which delivers an immersive environment [17–19].

4.7 Integrating AR/VR with digital twins

Incorporating AR/VR with digital twins can enhance the capability of these technologies to provide a comprehensive understanding of data and generate effective solutions. By using data visualization tools, stakeholders can gain valuable insights that might not be visible through traditional methods. The use of digital twins in conjunction with AR/VR can help identify areas that require improvement and enable employees to save time and reduce errors. The combination of IoT, AR, VR, and digital twins can be used across various industries to analyze data collected by sensors, turning it into valuable information for visualization. The potential benefits of integrating these technologies are vast and can result in long-lasting impacts.

4.7.1 Case study on Metaverse

Metaverse is a web of 3D digital world that is not limited to virtual gaming and comprises social media, web3, augmented reality, blockchain, virtual reality, digital twins, and many more technologies. VR and AR have a much broader role in delivering an immersive environment to consumers and thus act as a crucial element in the metaverse. End consumers can interact with others within the immersive surrounding and reduce the distance globally in the digital mode [20]. Users can take charge of their virtual avatars entirely by using VR headsets along with wireless body sensors, making users more immersed on the platform.

Tommy Hilfiger, one of the most popular fashion brands, provides their customers with a 360-degree complete view of fashion events by utilizing in-store VR headsets [7]. A shoe company, Toms, deployed VR to organize an online donation drive. Newly bought shoes are contributed to kids in need, and workers then travel across the world to donate the shoes directly to kids. Virtual headsets are made available in-store, and consumers can digitally interact with receivers, witness their lifestyle situation, and even witness their reactions upon receiving the shoes.

Augmented Reality allows end users to have a live panorama of real surroundings and expand it with virtual sensors such as audio, graphics, and videos. AR smart applications give consumers the opportunity to visualize the real world and make digital amendments to it through smart wearables and smart mobiles. For instance, the AR-based smart mobile application termed as IKEA Place provides customers with the opportunity to keep IKEA furniture in their house and visualize the view of furniture in their house. Consumers can click the images of the furniture displayed in their homes and share them with their loved ones [8]. Users can purchase the goods they are looking for through the IKEA website. The smart application by BMW on AR enables their potential customers to map out BMW 7 series car as per their preference. Users can wander around the digital vehicle and come across multiple information regarding the car. Based on that, users can get an idea about the purchase.

Organizations are developing digital twins of their automation plants to recognize mishaps in their production. Shopping virtually is usually complex because users are not entirely sure about the items and rely on images and consumer feedback. Due to this challenge, the e-commerce sector delivers a digital storefront where users can have visualization even after staying at their home [10]. Live updating of items that can be accessed is done. Metaverse builds the digital surrounding and magnifies reality, whereas digital twins can develop an exact duplication of the physical environment. Metaverse with Digital Twins aids in constructing digital architecture of real vehicles that have also started. People associated with the manufacturing sector can transmit data in real-time to the vehicles built in a digital surrounding, and performance of the vehicle can be evaluated. Developers can learn how a specific vehicle reacts in various conditions [9]. Cities can be created inside the metaverse by deploying digital twins to sketch the area, optimizing resources, and users can visualize how a specific area allocated for the city can be implemented efficiently [9].

4.8 Implementation

We have used three different files to create cloud-based digital twins on Azure Digital Twin Explorer by modeling individual entities as digital twins [23]. Real-time data about the environment is obtained using a knowledge graph. The "Floor.json" and "Room.json" files describe a building's floor and specific rooms, respectively. The JSON file detailing the graph and relationships between floor and room twins for a specific building is shown in Figure 4.4.

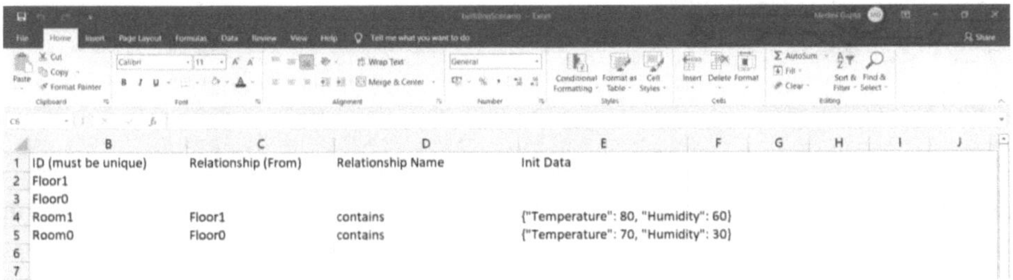

Figure 4.4: Building a scenario file.

To create the digital twins, we first instantiate them and then connect them to Azure's digital twin explorer. The digital twin model is defined in DTDL (Digital Twin Definition Language), as shown in Figure 4.5.

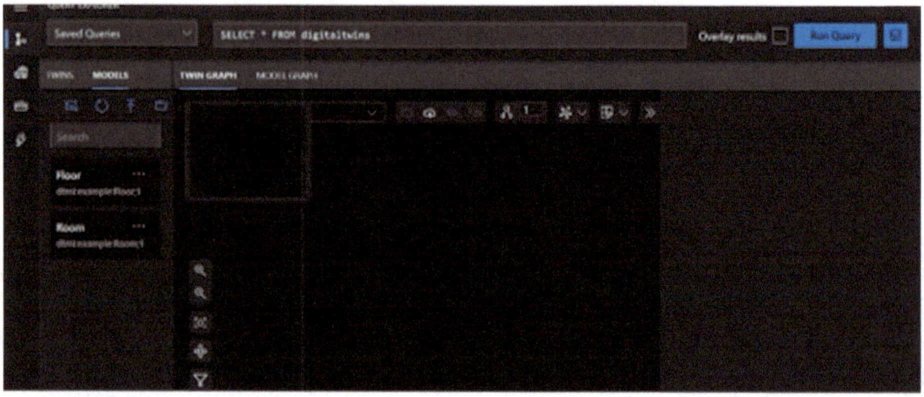

Figure 4.5: A json model being uploaded.

Navigate to both the files to view model details. Figure 4.6 and Figure 4.7 displays the complete information about json files being used.

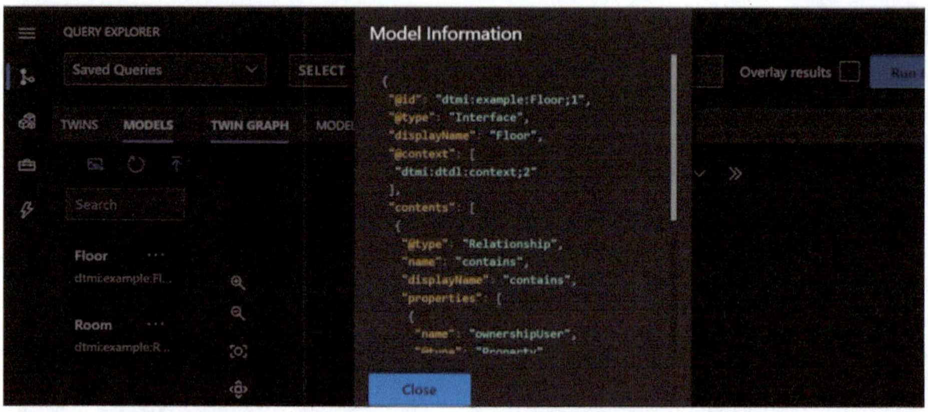

Figure 4.6: Floor,json details.

Import the BuildingScenario Excel file that contains details about a sample graph through which digital twins will be created. The graph consists of two rooms and floors each in a building. Figure 4.8 shows the graph preview. Figure 4.9 represents the successful import of twin models.

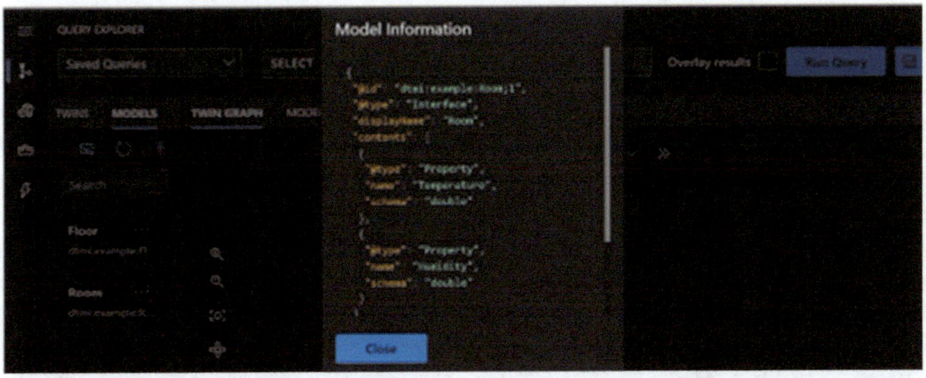

Figure 4.7: Room,json details.

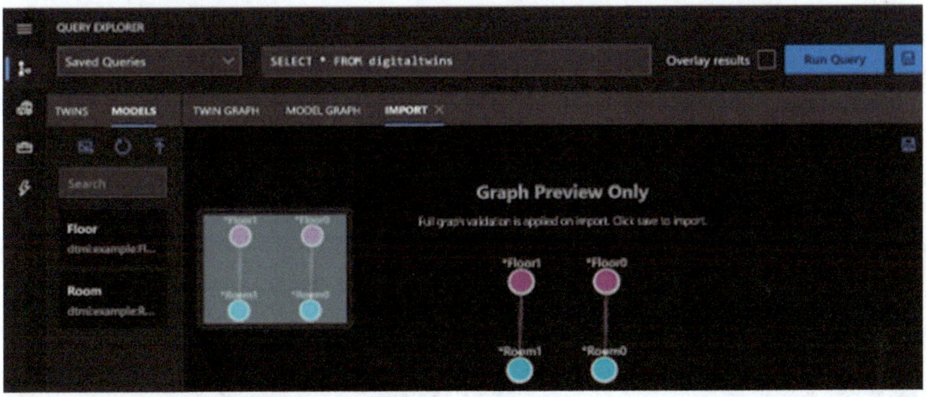

Figure 4.8: Graph preview.

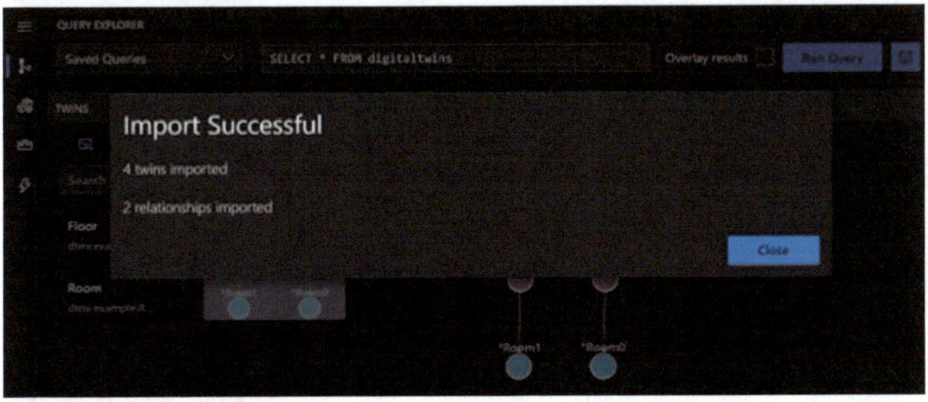

Figure 4.9: 4 twins imported.

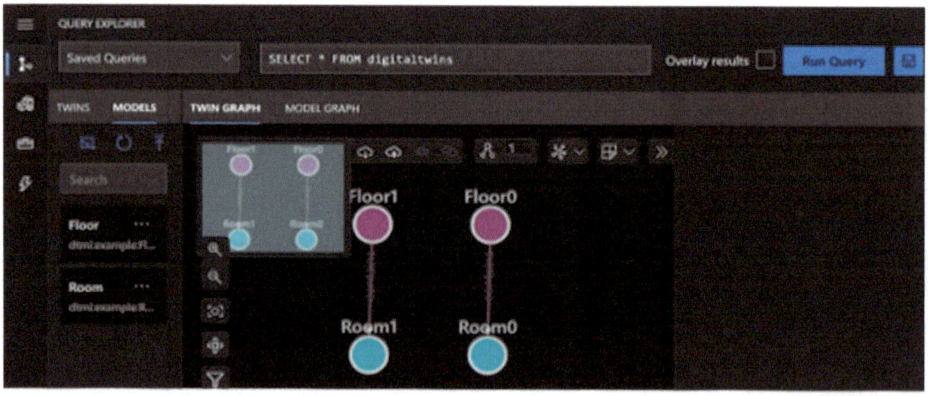

Figure 4.10: Twin graph panel.

Use the "Run Query" option to execute the query. Figure 4.10 shows the query statement that is used to retrieve all the digital twin in a graphical format.

We can explore the graph by viewing the characteristics of each twin. Figure 4.11 represents the twin features of Floor0, whereas Figure 4.12 shows the twin characteristics of Room0. Similarly, Figure 4.13 and Figure 4.14 display the twin features of Floor1

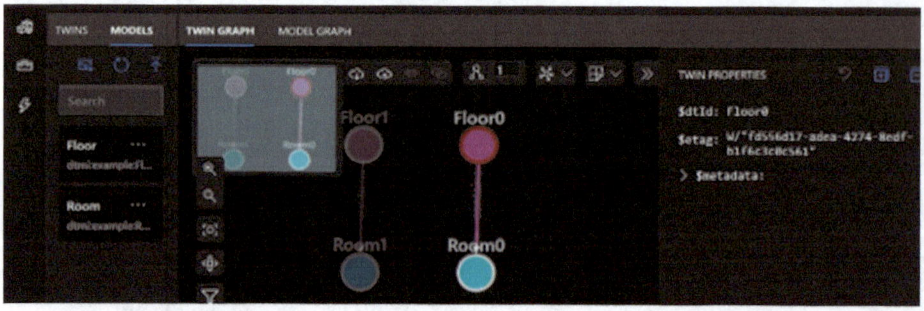

Figure 4.11: Twin properties of Floor0.

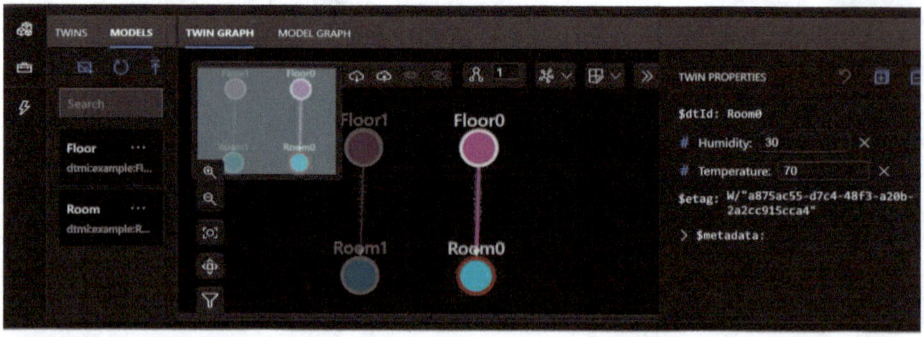

Figure 4.12: Twin properties of Room0.

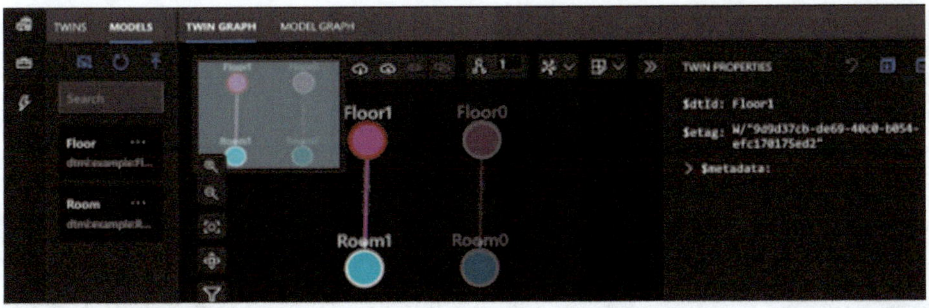

Figure 4.13: Twin features of Floor1.

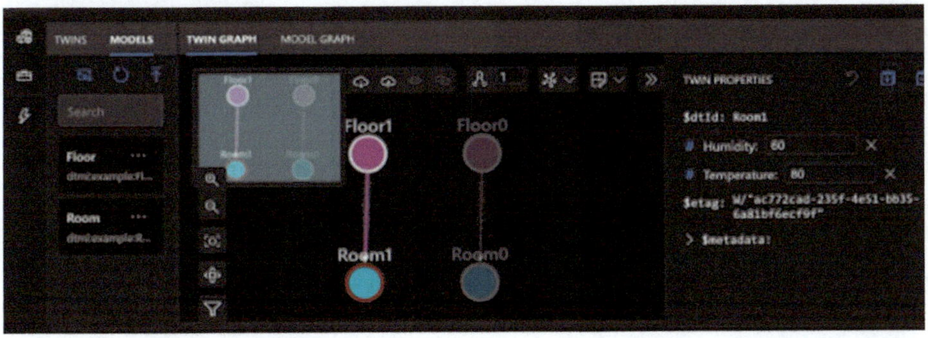

Figure 4.14: Twin features of Room1.

Figure 4.15: Query the twin.

and Room2, respectively. Room0 has a humidity level of 30 %, whereas Room1 has a humidity level of 80 %.

Azure digital twins have a query language from where users can state their queries and can get responses about their environment [24]. Figure 4.15 displays the query to know which twins in the environment have temperature greater than 78 °F.

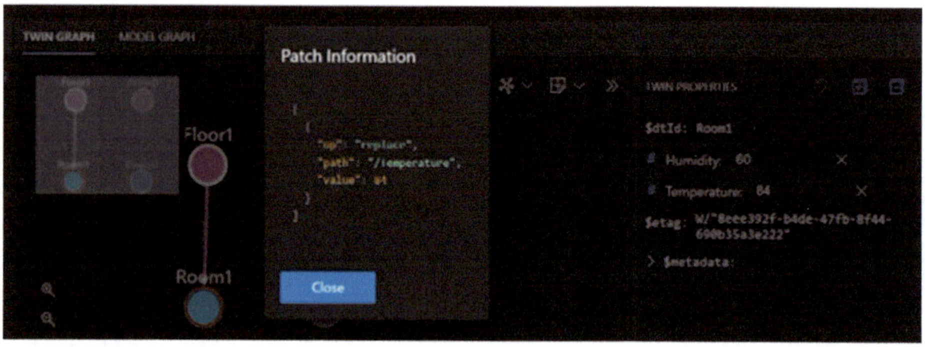

Figure 4.16: Values updated.

Digital twin model receives real time data gathered from IoT and accordingly update their characteristics as per the surrounding. Even the user can also manually update the details. Figure 4.16 shows the updated value of Room1. Earlier the temperature for Room1 was 80 °F whereas now the updated value is set to 84 °F.

4.9 Conclusion and future scope

The successful integration of AR, VR, and digital twins largely depends on further advancements in IT infrastructure. These technologies consume huge amounts of data and require significant investment in analytics and storage. To fully realize the benefits of these technologies, organizations must invest with a clear understanding of the potential business outcomes. AR and VR have demonstrated their potential in training and maintenance, while digital twins are more beneficial for predictive and diagnostic purposes [10]. Incorporating these technologies into industrial automation can greatly increase efficiency and enable real-time updates [11]. The combination of digital twins, AR, and VR with metaverse technology opens up countless opportunities, including the ability to shop in a digital world where information is updated in real time. These technologies have immense potential in various sectors, including quality monitoring, resource handling, and tracking. With an increasing number of organizations adopting these technologies, we can expect even more precise predictions and a closer merging of human and digital environments.

Bibliography

[1] E. Bottani and G. Vignali, "Augmented Reality Technology in the Manufacturing Industry: A Review of the Last Decade," IISE Transactions, vol. 51, no. 3, pp. 284–310, 2019, https://doi.org/10.1080/24725854.2018.1493244 (accessed September 2022).

[2] W. Tao, Z.-H. Lai, M. C. Leu, and Z. Yin, "Manufacturing Assembly Simulations in Virtual and Augmented Reality," Augmented, Virtual, and Mixed Reality Applications in Advanced Manufacturing., 2019.
[3] Å. Fast-Berglund, L. Gong, and D. Li, "Testing and Validating Extended Reality (xR) Technologies in Manufacturing," Procedia Manufacturing, vol. 25, pp. 31–38, 2018.
[4] A. Sharma, E. Kosasih, J. Zhang, A. Brintrup, and A. Calinescu. "Digital Twins: State of the Art Theory and Practice, Challenges, and Open Research Questions," Journal of Industrial Information Integration, 100383, 2022.
[5] A. Costantini, G. Di Modica, J. C. Ahouangonou, D. C. Duma, B. Martelli, M. Galletti, M. Antonacci, et al., "IoTwins: Toward Implementation of Distributed Digital Twins in Industry 4.0 Settings," Computers, vol. 11, no. 5, 67, 2022.
[6] J. C. Aurich, M. Glatt, A. Ebert, C. Siedler, and P. Webe, "Engineering Changes in Manufacturing Systems Supported by AR/VR Collaboration," Procedia CIRP, vol. 96, pp. 307–312, 2021.
[7] M. Gupta, S. Tanwar, A. Rana, and H. Walia, "Smart Healthcare Monitoring System Using Wireless Body Area Network," In 9th International Conference on Reliability, Infocom Technologies and Optimization (Trends and Future Directions) (ICRITO), 2021, pp. 1–5, 2021, https://doi.org/10.1109/ICRITO51393.2021.9596360.
[8] L. Kakkar, D. Gupta, S. Saxena, and S. Tanwar, "IoT Architectures and Its Security: A Review," In Second International Conference on Information Management and Machine Intelligence, Lecture Notes in Networks and Systems, vol. 166, pp. 87–94, 2021.
[9] P. Datta, S. N. Panda, S. Tanwar, and R. K. Kaushal, "A Technical Review Report on Cyber Crimes in India," In 2020 International Conference on Emerging Smart Computing and Informatics (ESCI), Pune, India, 2020, ISBN 978-1-7281-5263-9, pp. 269–275, IEEE, 2020, https://doi.org/10.1109/ESCI48226.2020.9167567.
[10] S. Tanwar, T. Paul, K. Singh, A. Rana, and M. Joshi, "Classification and Impact of Cyber Threats in India: A Review," In 2020 8th International Conference on Reliability, Infocom Technologies and Optimization (Trends and Future Directions) (ICRITO), 4–5 June 2020, ISBN 978-1-7281-7016-9, pp. 129–135, IEEE, 2020.
[11] R. Dhanalakshmi, C. Dwaraka Mai, B. Latha, and N. Vijayaraghavan, "AR and VR in Manufacturing," In Futuristic Trends in Intelligent Manufacturing, pp. 171–183. Springer, Cham, 2021.
[12] P. Wang, X. Bai, M. Billinghurst, S. Zhang, X. Zhang, S. Wang, W. He, Y. Yan, and H. Ji, "AR/MR Remote Collaboration on Physical Tasks: A Review," Robotics and Computer-Integrated Manufacturing, vol. 72, 102071, 2021.
[13] S. M. Hasan, K. Lee, D. Moon, S. Kwon, S. Jinwoo, and S. Lee, "Augmented Reality and Digital Twin System for Interaction With Construction Machinery," Journal of Asian Architecture and Building Engineering, vol. 21, no. 2, pp. 564–574, 2022.
[14] Z. Zhu, C. Liu, and X. Xu, "Visualisation of the Digital Twin Data in Manufacturing by Using Augmented Reality," Procedia CIRP, vol. 81, pp. 898–903, 2019.
[15] P. Novák and J. Vyskočil, "Digitalized Automation Engineering of Industry 4.0 Production Systems and Their Tight Cooperation with Digital Twins," Processes, vol. 10, no. 2, 404, 2022.
[16] J. Vrana. "Industrial Internet of Things, Digital Twins, and Cyber-Physical Loops for NDE 4.0," In Handbook of Nondestructive Evaluation 4.0, pp. 1–34, 2021.
[17] A. A. Malik and A. Bilberg, "Digital Twins of Human Robot Collaboration in a Production Setting," Procedia Manufacturing, vol. 17, pp. 278–285, 2018.
[18] Z. M. Cinar, A. A. Nuhu, Q. Zeeshan, and O. Korhan, "Digital Twins for Industry 4.0: A Review," In Global Joint Conference on Industrial Engineering and Its Application Areas, pp. 193–203. Springer, Cham, 2019.
[19] D. Mourtzis, V. Siatras, and J. Angelopoulos, "Real-Time Remote Maintenance Support Based on Augmented Reality (AR)," Applied Sciences, vol. 10, no. 5, p. 1855, 2020.

[20] S. Ke, F. Xiang, Z. Zhang, and Y. Zuo, "A Enhanced Interaction Framework Based on VR, AR and MR in Digital Twin," Procedia CIRP, vol. 83, pp. 753–758, 2019.
[21] Z. Zhu, C. Liu, and X. Xu, "Visualisation of the Digital Twin Data in Manufacturing by Using Augmented Reality," Procedia CIRP, vol. 81, pp. 898–903, 2019.
[22] D. Bamunuarachchi, D. Georgakopoulos, A. Banerjee, and P. P. Jayaraman, "Digital Twins Supporting Efficient Digital Industrial Transformation," Sensors, vol. 21, no. 20, p. 6829, 2021.
[23] https://azure.microsoft.com/en-in/get-started/azure-portal/ (accessed September 2022).
[24] https://azure.microsoft.com/en-us/services/digital-twins/ (accessed September 2022).
[25] A. Nayyar, B. Mahapatra D. Le, and G. Suseendran, "Virtual Reality (VR) & Augmented Reality (AR) Technologies for Tourism and Hospitality Industry," International Journal of Engineering and Technology, vol. 7, no. 2.21, pp. 156–160, 2018.
[26] D. Zingsheim, P. Stotko, S. Krumpen, M. Weinmann, and R. Klein, "Collaborative VR-Based 3D Labelling of Live-Captured Scenes by Remote Users," IEEE Computer Graphics and Applications, vol. 41, no. 4, pp. 90–98, 2021.
[27] N. Elmqaddem, "Augmented Reality and Virtual Reality in Education. Myth or Reality?" International Journal: Emerging Technologies in Learning, vol. 14, no. 3, 2019.

Rohan Mulay, Sourabh Singh Verma, and Harish Sharma

5 Methodical study and advancement in AR/VR applications in present and future technologies

Abstract: Communication has been essential for humanity's survival since its inception. Five hundred years ago, letters sent through messengers and pigeons were the means of communication with people living miles away from us. Over time, we evolved our ways of communicating, from hearing the voice of the person we're talking to via telephones to video calls that pushed the boundaries of communication and interaction. Augmented and Virtual Reality have the power to engage individuals in ways never before imagined. This chapter explores the sectors in which Augmented and Virtual Reality are directly impacting communication and how these changes can make our lives easier. Topics covered include the future of AR/VR, AR/VR applications, the AR/VR metaverse, human-computer interaction, and haptic technology.

5.1 Education

As our civilization moves forward and with companies even betting on setting colonies on other planets, the evolution of communication becomes even more important [1]. Although, online education has made every teacher on the web accessible to every student, this incredible accomplishment lags in one small point. AR and VR will make education a more immersive experience than before. This means a higher form of online education. We just can't get the same feeling as if we were in a classroom in the existing online education system [2]. This is where web 3.0 comes in; a student in India can learn from an instructor in the United States while experiencing the same classroom environment. Right now, it costs around 2 crore Indian Rupees to do MBA Degree from an Ivy League college like Harvard Business School in two years. More than half of the expenses are living expenses. With AR and VR, all of those living expenses will be eliminated.

AR/VR as an educational device is not really a new idea. Yet, immersive learning has as of late progressed from limited scope trial and emerged into a multimillion-dollar market with quickly developing use. Two classrooms across the nation use AR/VR for virtual field trips, science tests, and vivid re-enactments, and that's only the tip of the iceberg. Many vital interactions are possible with cell phones and advanced headphones. They are improving in quality while decreasing in cost. The innovations required to make and obtain vibrant content are also becoming easier to use and more economical.

Rohan Mulay, Sourabh Singh Verma, Harish Sharma, SCIT, Manipal University Jaipur, Jaipur, India, e-mails: mr.rohan99@gmail.com, ssverma80@gmail.com, harish27j@gmail.com

https://doi.org/10.1515/9783110785234-005

This paper explores the current state and likely commitments of AR/VR in education, as well as an assessment of the arrangements across subjects and learning levels that are laying the groundwork for the future's vibrant study halls.

Math Ninja AR makes learning math fun for younger children by incorporating physical movements such as stretching, bending, squatting, crawling, reaching, and peering to find the answer through the many levels inside the app to help stimulate a child's memory and help them retain information [3]. AR Pianist [4] is an interactive, modern way for students to learn how to play the piano, with features in the app such as providing them with a virtual piano if they don't have one, famous virtual pianists that can play songs right in front of them, the ability to slow down songs, learn songs with interactive sheet music, live feedback, and more. Animal Safari 3D [5] is a way for students to learn about animals visually. Animal Safari features the ability to see 1:1 scale animals in 3D, hear live animal noises, move and rotate any animal you want, feed, and watch your animal eat. Students can also pull up any animal's information card and learn about its habitat, diet, etc.

5.2 The future of AR/VR in higher education

Across the establishments of advanced education, there has been an increase in the arrangement of AR/VR for upgraded learning. Schools and universities are making foundations and committing assets to incorporate their educational plans with these advances. AR/VR innovation isn't economical and requires exceptional arrangements between the tech and the educational programs to convey desired learning results. This might hinder a few establishments to consolidate AR/VR in their schooling conveyance frameworks [6].

More noteworthy, greater awareness should be raised through studios and instructional exercises to persuade the teachers that the advantages of AR/VR far outweigh the expenses. The innovation can be demonstrated to the local community through special featuring AR/VR. Additionally, as the economies of scale come into play and the cost of AR/VR is smoothed out, innovations will continue to enhance the appeals of providing students with a real-file learning experience. With these advancements in educational technology, the future of higher education looks bright [7].

CamToPlan [8, 9] can be a valuable tool for students in fields such as engineering and architecture students, as is allows them to virtually measure anything in front of them. Additionally, the app can be helpful in interior design, landscaping, and other related fields.

5.3 Gaming

The gaming industry is expected to be a major area of growth for AR and VR technologies. In India alone, the online gaming population has grown from 200 million in 2018 to 400 million by 2020, with this number only set to increase in the coming years. In the early 2000s, there was a dotcom bubble in India and the USA, where every other person was either building a website or using one. A similar situation is expected to repeat itself for online games in India.

Gamers can experience an immersive environment, feeling as if they are inside a game [10], which will transform the gaming experience. Again, this is an example of how communication is evolving. Imagine playing PUBG, Fortnite or Valorant in such an environment. Ad-hoc networks such as MANET [11–13] and VANET can also benefit from AR/VR technologies, providing visual information about nodes or vehicles, such as malicious nodes. The potential for a gaming metaverse is also being explored, as showcased in the movie Ready Player One, which highlights how the gaming universe is likely to evolve.

5.4 Real estate

AR/VR technology is expected to have a significant impact on the real estate industry, particularly in the realm of augmented experience home visits. These involve using 360° cameras to film a home and create a seamless, immersive 3D experience for the viewer, allowing them to feel as if they are physically present in the property [14, 15]. A VR home visit allows clients to take measurements and learn about various aspects inside the home, but it its primary purpose is to measure the emotional appeal or "feel" that potential buyers have towards a property. This technology adds a new dimension to the traditional use of photographs and videos in showcasing a property's charm, allowing potential buyers to get a better sense of its emotional appeal.

In addition, AR technology can be used to place 3D models of furniture and other objects in a real environment, allowing clients to visualize what a room would look like furnished. (An excellent example of this is animated films like Toy Story). The construction business is one of the most established in the world; there has been significant progress in the tools we use to improve our built environment. Over the past decade, numerous advances have been introduced into the industry, including but not limited to, Building Information Modeling (BIM), 3D printers, Drones, and Virtual Reality (VR) [14]. One technology that has generated a lot of buzz due to its potential is Augmented Reality (AR). The potential of AR to become a significant tool in the building, engineering, and construction industry is enormous and in constant development.

Augmented Reality utilizes advanced camera and sensor technology to create an enhanced version of the real world by adding digital visual elements, sound, or other sensory stimuli and presenting it in real-time [16]. While this technology has been popular in the gaming industry for a long time, it is still relatively new in the construction industry. However, AR has been making waves in the industry due to the wide range of applications it provides across the project lifecycle.

5.5 Manufacturing

AR technology can be used in manufacturing for digital designing and mapping, and 3D printing can reduce labor costs while creating new opportunities and jobs. By producing a visual depiction of manufacturing plants paired with digital data, AR can improve planning reliability, resulting in savings in time and cost.

Computer vision can evaluate the film and determine how virtual goods should be placed. Further improved calculations can be used to simulate shadows, impeding, and kinematics, while AR PC vision is limited to a general location over a reference mark. The end result is a seamless blend presented to the client.

A GPS location is used as a trigger to match the current environment with AR data stored as a focal point. GPS, accelerometers, direction sensors, and barometric sensors as well as tourist spots, for example, road lines, structures, and the horizon can likewise provide direction [17, 18].

Dynamic expansion is another example of AR technology, which responds to changes in an object's viewpoint, such as the face, to apply and modify various types of beauty care products or eyeglasses using the facial picture as a perspective. This type of AR uses motion to determine the appropriate level of rising for the article based on the model's size.

If VR were used, manufacturing worker movements would be hindered, as their true interaction would be blocked with the headset on. AR devices, on the other hand, make worker movements more natural and seamless. [19].

The development can be used to assess a range of changes, detect unsafe working circumstances, or even visualize a finished product. It can show more than just automated text, graphics, or content, such as overlaying messages, subtleties, and information pertinent to the worker's ongoing task. For example, a worker can see a piece of equipment that is hot, indicating that it is dangerous to touch with bare hands. A manager can monitor everything that is happening around them, including the location of colleagues, equipment malfunctions, and production line issues.

In conclusion, security is a major concern. New delegates can be educated, trained, and protected consistently with the correct AR apps without wasting significant resources. When working with large, sophisticated machinery or hazardous materials, being prepared and trained can be expensive.

Skilled professionals can use Augmented Reality to check a related system that tells them exactly where items and commodities are located. They may use the AR structure to double-check critical information, which would assist in fulfilling the request. The workers can then retrieve the item and return it to its original location.

A support group may use Augmented Reality to identify exactly what equipment and supplies need to be updated, as well as any potential difficulties [20]. The same system may indicate their maintenance schedules, the last time they received assistance, potential areas of concern, and much more. AR can abstract conceptual principles from the support process, allowing for faster responses and fixes.

Traditional planning and prototyping can be quite extensive. Before anything is handed to actual creation and assembly, there can be continuous change communications and potentially several revisions to the product which can benefit from using AR. AR can speed this cycle by enhancing task visualization and improving collaboration and communication between the project team [21]. An organization's chief can use AR to see the actual product being designed and developed in real-time. The leader can provide feedback, insights, and perspectives that can help manage the constant changes during the design process.

5.6 Online dating

Right now, apps like Tinder, Omegle and Bumble allow individuals to meet online first and then meet in the real world if they like each other. While this has been successful, one aspect is that it is highly superficial, making it difficult to determine how they are in real life. AR will make this experience more immersive than ever. You might go out for coffee with someone who is in one city while you are in another, and the interesting part will be that you would feel as if the person is sitting right in front of you.

Innovation has changed how find a partner, and we know that the way people meet, connect, and fall in love will continue to be re-imagined by advancements in technology. Take virtual reality, for example.

One of the most significant advantages of VR is the possibility of remote and more immersive engagement than other media. We have previously discussed how people interact in virtual environments, but what about meeting your future partner in VR?

A virtual reality date would be a more serious and personal experience than existing online dating. When you go from a screen to VR, you get "presence," which is the impression of an actual person in the virtual place. Studies demonstrate that using technologies like Skype and Facetime fosters more trust and fulfillment when compared to merely conversing on the phone. It follows that VR dating will accomplish this to a much more remarkable degree.

Imagine a date where you embark on a spacewalk together or a romantic stroll around Paris while being serenaded by your favorite band. Augmented Reality becomes an acceptable approach for boosting interest by providing limitless locations and unique experiences for people to enjoy without spending a dollar.

There is a trend in the evolution of AR, which is a part of Web 3.0. Web 3.0 is a combination of AR, VR, and Crypto to make the user experience more immersive than before. Web 1.0 consisted of 'read-only sites' that allowed users to access information, while Web 2.0 was 'read and write sites' that allowed users to share information (e. g., Facebook, LinkedIn). However, in Web 2.0, users themselves became the product, and this lead to more and more companies monetizing user data. This centralization of power posed a problem, with only a few organizations controlling a lot of user data.

Web 3.0 aims to decentralize this power and give it back to the users. In this new paradigm, users are the creators and the ones who will benefit from creating and selling innovative digital products. The potential for innovation is sky-high, and new businesses in this sector have the opportunity to grow exponentially. For example, when a person sends money from one bank to another, they use the centralized servers of the financial providers. Banks act as intermediaries to carry out the transaction, and the customer must provide all the necessary data and rely on the bank to execute the transaction accurately. The bank charges a fee for this service. This is Web 2.0 banking.

In contrast, in Web 3.0, a person can send their transaction through a decentralized blockchain like the Bitcoin blockchain, which verifies the accuracy of the transaction using math and processing power. Banks are not always necessary as intermediaries in this case. Additionally, the user has control over their data, and since no centralized party is making money from the transaction, there are no fees to pay.

Ultimately, Web 3.0 aims to return data ownership and control to the user. Virtual Reality may play a role in the future of dating and love, thanks to its ability to engage people through multiple senses and allow them to converse from the safety of their own space. In a virtual setting, individuals may be able to see, hear, and even feel their partners.

5.7 Metaverse

Metaverse is a three-dimensional online experience that contrasts with the modern two-dimensional online experience that is viewed on a screen. In the metaverse, users will be able to "walk" through the experience using headphones or glasses. It is uncertain whether there will be a single metaverse or several independent ones. However, the metaverse will offer a cutting-edge version of the internet made possible by Virtual or Augmented Reality technology.

Matthew Ball, a financial investor whose thoughts on the metaverse have influenced Mark Zuckerberg, refers to the metaverse as a "replacement state to the flexible web" and a "platform for human leisure, activity, and presence at large".

A mirror world is a carefully delivered variant of this present reality where there are digital twins of genuine individuals, places, and things. The concept of mirror worlds appears often in science fiction, such as in Netflix's Stranger Things, The Matrix film series, and Ready Player One, a novel, and film. The metaverse could be a mirror world that closely reflects the real world, or it could be a wholly constructed world that experienced in a video game.

Skeuomorphic design refers to virtual objects that closely resemble real ones. Although the metaverse may be closely tied to the physics and designs of human life, it doesn't have to be identical to reality.

A digital twin is a virtual representation of a real object or structure. The concept was coined by David Gelernter in his 1991 book Mirror Worlds. NASA used digital twin technology for the first time in 2010 to simulate the interior of space capsules. Microsoft has emphasized the importance of the sophisticated digital twin technology in building the metaverse.

In a virtual environment, your avatar represents the user's persona. This digital representation of oneself can take on various forms from cartoonish (as Snapchat's Bitmoji and Apple's Memoji) to highly imaginative (such as in the case of Fortnite's "skins").

5.8 Social media

AR empowers digital information to be superimposed and integrated into our actual physical environment. With many of us now at home during a global pandemic, AR is a tool that can help us transform our immediate surroundings into learning, work, and entertainment spaces.

AR can assist with exploring the world beyond our physical limitations, from taking a virtual safari with 3D animals in our living room using Google's AR search on our smartphones to collaborating with avatars of remote colleagues as if we were in the same room using Spatial.

There are three things AR does particularly well: visualization, annotation, and narrating. There are examples in each of these areas that are both opportune in this era of COVID-19 and which can be built upon once social institutions, schools, and workplaces reopen their doors.

Virtual Reality refers to an immersive visual environment, which can include 360-degree videos, photos, or product demonstrations. It can also be enabled through devices like the HTC Vive or Oculus Quest. Retailers are currently using VR to either expand or enhance in-store experiences or digitally replicate the advantages of shopping in-person. Let's explore some of the ways in which retailers are currently utilizing VR, as well as some of the potential for even more advanced applications in the next few years.

By incorporating VR headsets in physical stores, retailers can provide customers with a unique way to view product options without taking up physical space in the store. For instance, in 2017, Audi used the Oculus Rift headset to enable customers to see their desired car in 3D and customize every aspect of it—from the paint color to

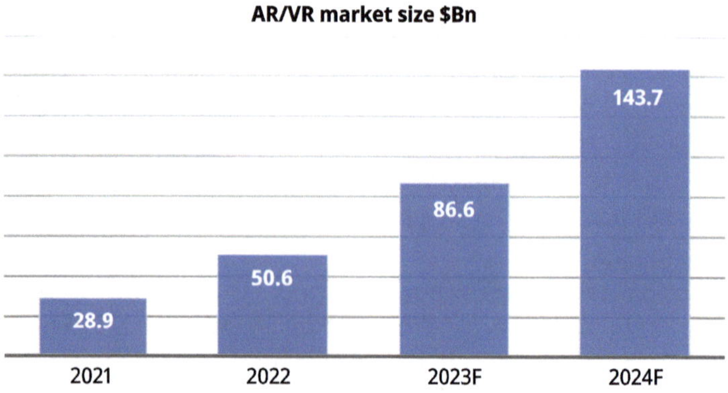

Figure 5.1: Market size of AR and VR (data source: netsubscribes).

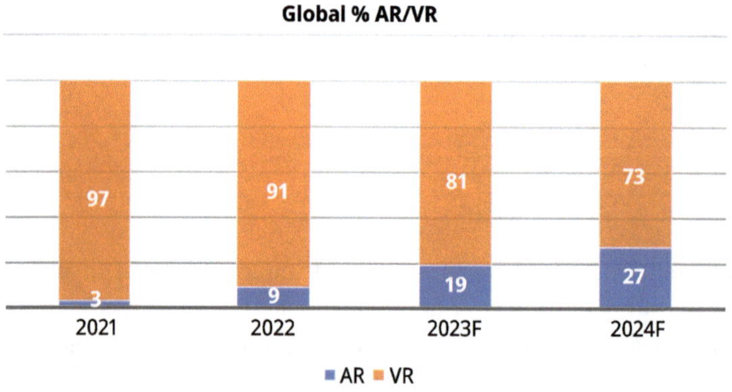

Figure 5.2: Proportion of AR and VR over the years (data source: netsubscribes).

the engine—in a more interactive way than simply selecting upgrades from a list on a PC screen. These immersive visualizations and interactions can help customers better understand what they want, ultimately enhancing the in-store shopping experience.

Despite the ongoing COVID-19 pandemic, VR can still be a useful tool in physical stores to provide customers with a memorable shopping experience as shown in Figures 5.1 and 5.2.

5.9 Medical and healthcare

Medical and healthcare are the most important applications of AR and VR. Training medical professionals in a real-life scenario can be quite expensive and challenging due to the unavailability of subjects. Here AR and VR can play a very important role by providing training to resident doctors and nurses. Moreover, even a normal person can be trained to provide emergency medical services like CPR (Cardio-Pulmonary Resuscitation) or ERS (emergency response service) using these technologies. Blum et al. have taken the first steps towards developing a Superman-like X-ray vision by using a brain-computer interface (BCI) device and a gaze tracker are to control the AR visualization.

The Complete Anatomy 2021 app is a modern way to learn about the human body by providing up-close 3D looks at various anywhere body parts like the heart, nerves, and muscles.

5.10 Space Research

Using AR and VR techniques, we can create simulated environments for robots and space rover to explore other planets or spaces. It can reduce the chances of rover failure and save large amounts of money invested in space missions. AR and VR techniques are also used in training pilots and astronauts for spacecraft and space stations.

The first use of AR on the International Space Station (ISS), a set of high-tech goggles called Sidekick, provided hands-free assistance to crew members using high-definition holograms that show 3D schematics or diagrams of physical objects while they completed tasks. Additionally, the goggles included video teleconference capability that allowed the crew to receive direct support from flight control, payload developers, or other experts.

The T2 Augmented Reality (T2AR) project demonstrates how station crew members can inspect and maintain scientific and exercise equipment critical to maintaining crew health and achieving research goals without assistance from ground teams. T2AR is the first in-space operational use of the HoloLens in combination with custom-built AR software, which enables astronauts to perform unassisted maintenance and inspections on critical crew support hardware. The system also provides additional information, such as instructional videos and system overlays, to assist with performing the procedures.

The ISS Experience is an immersive VR series that documents over multiple months various crew activities, from conducting science experiments aboard the station to performing spacewalks. The series uses specialized 360-degree cameras designed to operate in space to transport viewers to low-Earth orbit and create the sensation of being an astronaut on a mission. It also provides audiences on Earth a better understanding of the challenges of adapting to life in space, the work and science that take place, and the human interactions between astronauts. This could spark ideas for research or programs to improve conditions for crew members on future missions and inspire future microgravity research that benefits people on Earth.

5.11 Virtual land

Compared to gold, equities, mutual funds, bonds, and other investments, land has historically been an exceptional investment. In the words of Mark Twain, “Buy land, they’re not making it anymore." Owning land or investing in it has become a dream for many as the population grows and development reaches its zenith. Consider a scenario in which we inform you that you may claim land right now, and despite its insignificance, it is more lucrative than the current land industry.

Fuelled by cryptocurrency, the metaverse is undoubtedly the most sweltering fascination for investors looking to buy land. Many people may be surprised by the prospect of virtual lands, but for those who are familiar with games like Farmville, Clash of Clans, and The Sims, it is definitely nothing to laugh about. In fact, we've recently seen individuals and larger organizations offer large sums of money for a plot of real estate.

Let's discuss how to acquire land in the metaverse first. It's crucial to understand what the metaverse is. For the unenlightened, the metaverse is a virtual world without limits. Imagine a world where you can shop, buy, and explore—and possibly do everything that you do in real life. It ought to be noticed that it's anything but a substitution for the real world; it is updating your perspective of objects in digital space.

Land in the metaverse is a pricey business, just like in real life, where prices are determined by location, population, and the supply-demand ratio. When there is a surge in demand for a plot in the metaverse in a particular place, the costs rise accordingly. Some plots in the metaverse may cost more than $4 million, while others are available at reasonable prices. The unique feature of the virtual world is that ordinary people can purchase direct plots or even a virtual holiday island in the metaverse. Some compare purchasing real estate in the metaverse to purchasing real estate in Manhattan in the 1940s. Not every plot is as expensive, and some start at several hundred dollars. However, everyone should strive to be a part of the better regions in popular metaverses like the Sandbox and Decentraland, just like in real life.

Sandbox (SAND), Axie Infinity (AXS), Decentraland (MANA), Enjin (ENJ), and other notable metaverse initiatives have attracted land investments. These initiatives rely on the Ethereum blockchain and use digital currency for transactions in the metaverse. Apart from the aforementioned, gaming groups such as Atari and Roblox are leading various metaverse efforts.

While there are many plots available on several metaverses, the most popular Decentraland metaverse project is divided into 90,601 plots, and the Sandbox is divided into 166,464 parcels.

This raises a significant question: how could you purchase land in the metaverse, if it costs just as much as real-life property? The straightforward response is that the return on investment from speculation is a lot higher compared to real-life plots, with ROI sometimes reaching as high as 1000 % in a shorter timeframe.

Furthermore, blockchain technology makes it challenging for any land scams to occur. Blockchain technology is essentially a distributed ledger, meaning that every transaction is recorded, bringing more transparency to transactions. This limits the possibility of land scams like forced undoing, selling without authorization, false promises, and delays in ownership transfer.

5.12 Conclusion

After reviewing the advancements in AR and VR in various growth sectors, it is evident that these technologies are making communication more immersive than before. As human interaction has evolved from sending messages via pigeons to feeling real-time interactions with people on the other side of the planet, it is clear that this will be a key factor in our future expansion as a species.

As we have progressed from pigeons to messengers (due to the development of transportation) kingdoms started expanding beyond their continents (such as the British Empire). With the development of telephones that expansion has become easier and more viable. With Facebook and WhatsApp communication has become even more widespread and accessible to a larger audience. The question of forming colonies on other planets and thereby the expansion of mankind arises. We think AR and VR may be the stepping stones to a much more sophisticated level of communication, possibly utilizing holograms and a more interactive and understanding AI.

Bibliography

[1] L. B. Jaloza, "Inside Facebook Reality Labs: The Next Era of Human-Computer Interaction," Tech at Meta, 09-Mar-2021, (Online). Available: https://tech.fb.com/ar-vr/2021/03/inside-facebook-reality-labs-the-next-era-of-human-computer-interaction/ (accessed: 26.07.2022).

[2] H. Sharma, S. S. Verma, V. Sharma, and A. Prasad, "Impact and Challenges for the Indian Education System Due to the COVID-19 Pandemic, Impacts and Implications of COVID-19," An Analytical and Empirical Study, pp. 63–85, 2021.

[3] "12 Augmented Reality Apps for the Classroom," TeachThought, 18.08.2016. (Online). Available: https://www.teachthought.com/technology/augmented-reality-apps (accessed: 26.07.2022).

[4] D. Deahl, "AR Pianist App Is Fun to Watch, but That's About It," The Verge, 13.02.2020. (Online). Available: https://www.theverge.com/2020/2/13/21136248/ar-pianist-massive-technologies-virtual-musicians-pianos (accessed: 26.07.2022).

[5] "Animal Safari AR," (Online). Available: https://play.google.com/store/apps/details?id=io.lightup.safari&hl=en_IN&gl=US (accessed: 26.07.2022).

[6] M. Billinghurst and A. Dunser, "Augmented Reality in the Classroom," Computer, vol. 45, no. 7, pp. 56–63, 2012.

[7] S. C.-Y. Yuen, G. Yaoyuneyong, and E. Johnson, "Augmented Reality: An Overview and Five Directions for AR in Education," Journal of Educational Technology Development and Exchange, vol. 4, no. 1, Jun. 2011.

[8] "Cam to Plan -the Best Mobile App to Make Your House Room Plan Using Your Smartphone. Download the Free APK File of This Mobile App Here," Cadbull. (Online). Available: https://cadbull.com/detail/165457/Cam-to-plan--the-best-mobile-app-to-make-your-house-room-plan-using-your-smartphone.Download-the-free-APK-file-of-this-mobile-app-here (accessed: 30.07.2022).

[9] "CamToPlan – AR Measurement / Tape Measure." (Online). Available: https://play.google.com/store/apps/details?id=com.tasmanic.camtoplanfree&hl=en_IN&gl=US (accessed: 3007.2022).

[10] R. Cavallaro, M. Hybinette, M. White, and T. Balch, "Augmenting Live Broadcast Sports With 3d Tracking Information," In IEEE MultiMedia, pp. 38–47, 2011.

[11] S. S. Verma, S. K. Lanka, and R. B. Patel, "Precedence Based Preemption and Bandwidth Reservation Scheme in MANET," International Journal of Computer Science Issues (IJCSI), vol. 9, no. 6, p. 407, 2012.
[12] S. S. Verma, A. Kumar, and R. B. Patel, "QoS Oriented Dynamic Flow Preemption (DFP) in MANET," Journal of Information & Optimization Sciences, vol. 39, no. 1, pp. 183–193, 2018.
[13] R. Soni, A. K. Dahiya, and S. S. Verma, "Limiting Route Request Flooding Using Velocity Constraint in Multipath Routing Protocol," In A. Somani, S. Srivastava, A. Mundra and S. Rawat, editors, Proceedings of First International Conference on Smart System, Innovations and Computing. Smart Innovation, Systems and Technologies, vol. 79, Springer, Singapore, 2018. https://doi.org/10.1007/978-981-10-5828-8_12.
[14] V. P. Brenner, J. Haunert, and N. Ripperda, "The Geoscope – A Mixed-Reality System for Planning and Public Participation," In 25th Urban data management symposium, 2006.
[15] O. Hugues, P. Fuchs, and O. Nannipieri, "New Augmented Reality Taxonomy: Technologies and Features of Augmented Environment," In Handbook of Augmented Reality, pp. 47–63, Springer, 2011.
[16] R. L. Silva, P. S. Rodrigues, J. C. Oliveira, and G. Giraldi, "Augmented Reality for Scientific Visualization: Bringing Data Sets Inside the Real World," In Proc. of the 2004 Summer Computer Simulation Conference, pp. 520–525, Citeseer, 2004.
[17] J. M. Krisp, Geovisualization and Knowledge Discovery for Decision-Making in Ecological Network Planning, 2006.
[18] N. R. Hedley, M. Billinghurst, L. Postner, R. May, and H. Kato, "Explorations in the Use of Augmented Reality for Geographic Visualization," Presence (Camb.), vol. 11, no. 2, pp. 119–133, Apr. 2002.
[19] T. P. Caudell and D. W. Mizell, "Augmented Reality: an Application of Heads-up Display Technology to Manual Manufacturing Processes," In Proceedings of the Twenty-Fifth Hawaii International Conference on System Sciences, Kauai, HI, USA, 1992.
[20] G. Reinhart and C. Patron, "Integrating Augmented Reality in the Assembly Domain - Fundamentals, Benefits and Applications," CIRP Annals – Manufacturing Technology, vol. 52, no. 1, pp. 5–8, 2003.
[21] Y. Baillot, D. Brown, and S. Julier, "Authoring of Physical Models Using Mobile Computers," In Proceedings. Fifth International Symposium on, pp. 39–46. IEEE, 2001.

Sangeeta Borkakoty, Daisy Kalita, and Prithiraj Mahilary

6 Application of Augmented and Virtual Reality for data visualization and analysis from a 3D drone

Abstract: The use of immersive technologies such as Virtual, Augmented and Mixed Reality has increased the demand of unstructured 3D data, including point clouds, which are becoming essential for geospatial applications in various sectors, like urban planning and development, building information modeling, and natural and cultural asset documentation. In this paper, we present and open-source solution using Unity to integrate a 3D point cloud from date from a drone into a virtual reality environment using photogrammetry. Our solution provides a user interface and rendering technique for VR point cloud interaction and visualization, enhancing both rendering performance and visual quality. Users will be able to examine a 3D point cloud in detail and perform immersive exploration and inspection of sizable 3D point clouds on cutting-edge VR devices thanks to a set of interaction and locomotion mechanisms.

6.1 Introduction

Virtual Reality (VR) and Augmented Reality (AR) are two digital tools that either overlay visuals on the real environment (AR) [1–3] or give users a completely synthetic digital experience (VR) [13]. While products like games, social media filters, and gaming headsets have captivated people's interest worldwide, AR and VR also have numerous practical applications. Due to their ability to provide real-time data, engage consumers, and accurately replicate real-life experiences [5], AR and VR have immense potential to transform services and consumer interactions. Some potential uses of AR and VR include policing, emergency management, asset management, tourism, education, urban planning, and training.

With the advancement of remote and in-situ sensing technology, it is now possible to create extremely highly detailed digital models of real-world assets, locations, cities, or even entire countries in a cost-effective and time-efficient manner [4]. Large unstructured collections of 3D points—commonly referred to as 3D point clouds [6]—are the resulting data sets—offering a precise, thorough, and comprehensive digital representation of real-life items. In recent years, 3D point clouds have become a crucial data type for geospatial applications across various fields.

Sangeeta Borkakoty, Daisy Kalita, Prithiraj Mahilary, Department of Computer Science & Electronics, University of Science & Technology, Meghalaya, 793101, India, e-mail: s.borkakoty06@gmail.com

https://doi.org/10.1515/9783110785234-006

Virtual Reality devices such as Oculus Rift [8], Samsung Gear VR [9], HTC Vive [10], Google Daydream View [11], and Google Cardboard [12] have emerged in recent years, providing users with new ways of viewing and interacting with digital 3D material. These devices allow 3D point clouds to be viewed in an interactive manner, granting users the sensation of being physically present the recorded location [7].

In this paper, we present a rendering system that enables users to explore massive 3D point clouds in an immersive manner on VR devices. Our goal is to implement 3D objects or environments captured from the real world into a virtual workspace, providing users with detailed visualization, interaction and navigation capabilities.

Our objective is to enable users to interact with the presented data through methods like measuring areas and distances, as well as rotating and scaling. Additionally, we aim to incorporate various real-time both natural and artificial locomotion strategies. For this project, we have used open source VR development tools and software.

The key features of this application include:

- Data visualization techniques: storage, display and formats
- Natural interaction methods: marker-based, marker-less, hand control, and navigation, and optionally, gesture control navigation.
- Interaction with artificial digital content
- Ability to annotate and label places or objects within the 3D data.

6.2 Tools and technology

6.2.1 Data

The 3D model was shared by NESAC (North Eastern Space Applications Centre, Meghalaya), generated from 347 drone images captured by a senseFly drone sensor.

6.2.2 Point cloud

A point cloud is a large collection of individual points plotted in 3D space [14]. They are commonly generated by 3D scanners or photogrammetry software [15], which measure multiple points on the surfaces of objects.

6.2.3 Cloud compare

Cloud compare is a 3D point cloud processing software that can compare two 3D point clouds or a point cloud and a triangular mesh [16]. It works on an octree structure (a tree data structure in which each internal node has exactly eight children), which partitions a 3D space by recursively subdividing it into eight octants.

6.2.4 Unity engine

Unity [17] is a cross-platform game engine used to create real-time 3D projects in various areas such as games, animation, engineering, manufacturing, and construction. The Unity engine can be used to build both 3D and 2D applications, as well as interactive simulations (VR, AR, MARS).

6.2.5 XR interaction toolkit

We managed our target platform SDKs using Unity's XR Plugin Management package [18], and we added interactivity to our VR applications using Unity's XR Interaction Toolkit.

To build our AR experience [19] we used a component-based, high-level interaction system known as Unity's XR Interaction Toolkit. It offers a framework that allows Unity input events to access 3D and UI interactions. The system's main building block is composed of a set of basic *Interactor* and *Interactable* components, as well as an *Interaction Manager* that connects these two types of components. It also has elements that may be employed for movement and image creation. Block diagram of unity XR plug is given in Figure 6.1.

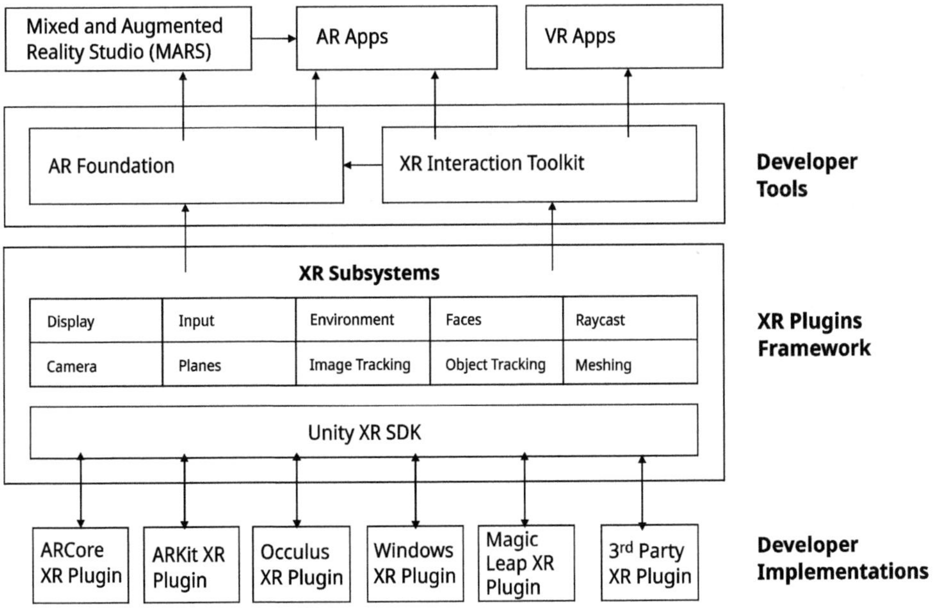

Figure 6.1: Block diagram of the Unity XR plug-in framework and how platform provider implementations interact with it. (Source: https://docs.unity3d.com/Manual/XRPluginArchitecture.html.)

6.2.6 Components

The XR Interaction Toolkit includes various types of components:
- **Interactors:** These components handle the hovering and selection of Interactable items. In each frame, the Interactor generates a list of valid targets over that can be hovered or selected.
- **Interactables:** These components are objects in the virtual scene that can be hovered over, selected or activated by an Interactor.
- **Interaction Manager:** Interactors and Interactables are connected by the Interaction Manager. It is responsible for causing state changes in the interaction between registered Interactors and Interactables. At least one Interaction Manager must be present in the set of loaded scenes for interactions to function properly.

6.2.7 States

Hover, *Select*, and *Activate* are the three basic states of the Interaction system. For various objects, these states can mean different things. The typical GUI concepts of mouse-over and mouse-down are similar to the states *Hover* and *Select*. *Activate* is a contextual command that is specific to XR.

Both an *Interactor* and *Interactable* are involved in these states, and both of them are notified when they enter or exit a state.
- **Hover:** An Interactable's state changes to Hover when it is a valid target for the Interactor. Although hovering over an object indicates a desire to interact with it, most of the time the object's behavior remains unchanged. However, it may produce a visual signal to indicate this state transition.
- **Select:** To enter the Select state, the user must perform an action such as pressing a button or trigger. The selecting Interactor is regarded as engaging with the Interactable when it is in the select state.
- **Activate:** An additional action that has an impact on the currently selected object is activation. It is usually assigned to a button or trigger. The user can now interact with the selected object in greater detail in this state.

6.2.8 AR Foundation

AR Foundation [20] is a package that enables the use of Augmented Reality platforms in multi-platform way within Unity. The AR package includes interaction components that can be used to implement any AR feature.

AR interaction is often driven by the AR Gesture Interactor component [21], which converts touch events into actions like tapping, dragging, and pinching. These gestures are then passed on to gesture Interactables, where they are converted into interactions.

Screen touches are transformed into motions using the AR Gesture Interactor component. Interactables receive the motions and process them before responding to the gesture event. An XR Origin or AR Origin must be present in the scene to enable the AR Gesture Interactor and its gesture recognizers to function.

6.2.9 Placing items using the AR Placement Interactables

To simplify the process of placing objects in the scene, we use the "AR Placement Interactable" component. Users can provide a placement prefab, and when a tap occurs, Unity places it on an AR plane. At the same time, Unity produces a ray cast directed at the plane. To enable further gesture interactions, the Prefab may also contain additional AR Interactables.

Following concepts are implemented in our AR environment system:

- **Device tracking:** keeps track of the device's position and orientation in physical space.
- **Surface detection using planes:** identifies horizontal and vertical surfaces.
- **2D image tracking:** detects and tracks 2D images.
- **Tracking 3D things:** locates 3D items.
- **Raycast:** searches for identified planes and feature points in the physical environment.

6.3 Materials

To fully interact with the VR system and experience an immersive exploration a VR device is required. The typical VR device consists of a headset and left and right controllers. The headset is generally configured with head and eye tracking features, while the controllers are configured with hand movement and gesture tracking systems.

VR devices consist of left and right controllers which come with several buttons. Each button has its own functionality and for interacting with the VR application.

6.4 Methodology

6.4.1 Loading

When data loads into the system, the necessary codes and textures are loaded into primary memory and processed to ensure a smooth interactive experience without lag or system slowdowns. This process involves reading data, typically graphics and environment data, from the hard disk and system memory and then sending it to the GPU.

6.4.2 Rendering

Rendering is the process of generating 2-dimensional or 3-dimensional images from a model using a computer program. Rendering involves a combination of geometry calculations, textures, surface treatments, the viewer's perspective, and lighting.

To render content in our scene, we use Unity's Built-in Render Pipeline. It performs a series of operations on the contents of a scene and then displays them on a screen. Rendering Pipeline consists of 3 phases:

APPLICATION ⇒ **GEOMETRY** ⇒ **RASTERIZATION**

- The CPU receives instructions for managing geometry and rasterization during the application phase. There are numerous CPU processes running in the background, including input, scene loop, audio, physics, and culling. Rasterization and geometry are impacted by this stage.
- Using the mesh data to generate 3D and 2D models, the system simulates the virtual world in those dimensions. This stage involves doing calculations regarding the camera location as well as the scaling, rotation, and alteration of the virtual environment.
- The virtual environment is processed numerous times through various filters throughout the rasterization step, and the outcome is displayed on the screen. Since our screens are 2D, the geometry (both 3D and 2D) will be drawn on them during the rasterization process.

The rendering processes are as below:

1. **Geometry**: The vertices array, normals array, triangle array, and UV array are examples of mesh data that are gathered here.
2. **Illumination**: The models are colored and lighted in this illumination step. Here we enhance the virtual world by using lighting effects. Inputs such as textures, normal maps, and others can be used to color virtual objects.
3. **The viewer's perspective:** Before the world is rendered on the screen, we must take into account the camera input, such as the field of view and the projection mode (orthographic or perspective), among other things.
4. **Clipping:** The process of removing objects that are outside the camera's field of view is called clipping.
5. **Screen space projection:** This technique involves displaying a 3D environment on a 2D screen.
6. **Post-processing effects** are those that are applied to the 2D image immediately before the final result is displayed on the screen.
7. **Display:** The final stage involves rendering our scene. The complexity of the data determines how long it takes to render a mesh.

6.4.3 Shaders

Shaders [22] are programs that modify the way materials appear on objects in a scene, by using a combination of texture UV coordinate information, object vertex, or pixel information, scene lighting information, and more in mathematical functions. Shaders are essential because they manipulate color information to apply any effect that we may want to use. Without proper shaders, the objects in the scene may not appear on the screen or may look visually deformed.

Unity comes with a number of built-in shaders that we can use in our project. Here we have used Unity's Standard (Specular Setup) Shader and during runtime, we have used Shader/Particles/Standard Unlit, Legacy Shaders/Particles/Multiply (Double), Legacy Shaders/VertexLit, Legacy Shaders/Particles/~Additive-Multiply, and Mobile/Bumped Specular. We have also use a point cloud shader to render point cloud data in our scene.

6.4.4 Locomotion

The set of locomotion methods included in the XR Interaction Toolkit package can be used for navigation within in a scene. These include:
- The user is represented by the XR Origin which the locomotion system manages to access.
- Teleportation device
- Snap Turns: Allow users to turn the rig at predetermined angles
- Continuous turn: The rig is smoothly rotated over time by a continuous turn provider
- Smooth movement: Allows users to smoothly move over time by a continuous movement provider.

6.4.5 Movement

In order for the human eye to detect natural movement in 3D, the rendering speed must be at least 24 frames per second. Realistic, natural movement can be challenging to achieve and requires many intricate computations. Unity, however, provides us with straightforward functions that we can employ to produce smooth movements.

To achieve a smooth movement in our scene, we have implemented following techniques:
- **Update function:** The update method is called every frame per second, which means it runs periodically while our scene is active. We use this method when we want to induce changes in time, such as movements. So, to move the user in our scene, we use the update method to change the position and rotation in each frame, adding dynamism to the scene.

- **Transform:** Every object in a scene has a transform function associated with it, which stores the position, rotation and dimensions and can be used to manipulate these parameters. By dynamically changing the values in the update method, we can induce movement in our objects.
- **Vector3 and translate:** Unity uses Vector3 to represent 3D vectors and points and to transfer 3D locations and directions. Additionally, it has routines for typical vector operations. In the Unity program, the translate function moves an object along a 3D axis at a speed of one unit per second while using the global coordinate system.
- **Speed value and Time.deltaTime:** The speed at which an object moves depends on the float value set and Time.deltaTime. Time.deltaTime is the completion time in seconds since the last frame, i. e., the time passed since rendering the last frame.

6.5 System testing

The system was tested in the following categories:

- **Directory management system module testing:** Ensuring that external files (.obj data) from a directory can be imported and loaded into the system.
  - Searching whether the mesh and texture files within the directory and importing them into the system.
  - Checking whether the system can read and load the mesh and texture files.
  - Verifying whether file paths are necessary for loading data.
- **Movement and locomotion system module testing:** Ensuring that the user can move freely in the world space using different options.
  - Checking whether the user can teleport into random locations using the teleport anchor.
  - Checking whether the user can move in the 3D axis using the directional arrow.
- **Marking and measurement module testing:** Ensuring that the user can mark around the world space as well as measure the distance between two points.
  - Checking whether the user can mark anywhere using the right controller.
  - Verifying whether dragging two points can show the updated distances in the world space.
- **Placing 3D objects module testing:** Ensuring that 3D objects can be placed in a scene.
  - Checking whether the object tray appears when clicking the desired button.
  - Checking whether the object appears in the world space while pressing the desired button.
  - Checking whether different 3D objects can be placed and removed in the world space.
- **Labeling 3D objects module testing:** Ensuring that objects can be labeled with a tag using a virtual keyboard.

  - Checking whether the virtual keyboard appears on activation.
  - Verifying whether all keys are showing the correct output on typing.
- **Shaders and camera view distance system module testing:** Ensuring that changing shaders takes effect on the scene.
  - Checking whether shader tray appears with multiple shader options.
  - Checking whether shader buttons are able to change the desired shaders.
  - Checking whether the user can change the view distance of a camera.
- **AR plane and image tracking system module testing:** Ensuring that our device can track any plane horizontally and vertical on the surfaces and also track image on a 2D reference object.
  - Checking whether the device camera can track plane surfaces.
  - Checking whether the device camera can track 2D reference objects.

6.6 Implementation

Setting up the scene and camera with XR Interaction Toolkit

First, we define the origin utilizing XR Origin, which represents the world-center space in an XR scene. In the Unity scene, the XR Origin adjusts objects and trackable features to their final position, orientation, and scale. A Camera Floor Offset Object, an Origin, and a Camera are defined.

Canvas and event system for user interface

The Canvas is the area within which all the UI elements should exist. The event system manages input, raycasting, and event dispatching. In a Unity scene, the *EventSystem* is responsible for processing and handling events.

Input system and scripting

The Input System in Unity implements a system to use input devices to control scene content. In-app elements can be programmed to respond to user input in different ways.

Scripting is a critical component of every Unity application. Scripts can be used for a variety of tasks, including the implementation of a bespoke AI system for scene characters as well as the creation of graphic effects and managing the physical behavior of objects.

Reading and loading OBJ and MTL data

A 3-dimensional object's 3D coordinates, polygonal faces, texture maps, and other object data are all included in an OBJ file. OBJ files come in binary (.obj) and ASCII (.obj) formats (.mod). However, they do not include descriptions of facial colors. There is no visual data in OBJ files. However, by referencing the MTL file, this information may be added to the objects in the OBJ file.

MTL (Material Template Library) is a widely used ASCII file format. It defines how an object in an OBJ file reflects light.

3D object materials with MTL filename extension generally contain ASCII text that defines how colors, textures, illumination, and reflection map of individual materials are applied to the surfaces and vertices of 3D objects in a model or a scene.

OBJ files can be opened using any 3D editing software. When an OBJ file is imported in the 3D editing software, the system will detect the type of format along with the reference MTL and texture file automatically and load the model. The loading time of a model depends on the size of the OBJ file and corresponding visual details such as texture complexity and level of details of a model and also the algorithm used by the software.

In our project we aim to import models during runtime, which requires the use our own algorithm for loading the 3D model. The process we follow is outlined below:

- Firstly, we need to set the file path of the OBJ and the texture that we need to load in our system. To retrieve the file path we use the system's input and output file stream.
- Secondly, we need to read the input of the OBJ data and get vertices, normals and UVs in a list to add material properties to each shape and geometry.
- Lastly, we need a texture loading function to load texture from the file path and add it to the material. After successfully implementing it can load OBJ, MTL and textures in decent import speed around 750k triangles in about 10 seconds; it works in editor mode and during runtime; it loads triangles, quads and ngons.

6.7 The VR UI application

File manager interface Upon entering the system, the first scene displayed will be the main menu where the user can load a demo scene or import an external file to load a scene. When the user clicks the Import button, the file explorer interface will appear and the user can choose an .obj file and .jpg file from the directory. After clicking the Load button the user is transported to another scene where the file has been loaded.

Loaded scene After a scene is loaded, the user is positioned in a specific location. By pressing the Menu button on a controller, a UI with different interactions is displayed. Each button performs different actions and helps the user to interact with the system. The UI can be turned on and off anytime with the Menu button on a controller.

Movement and locomotion UI The user can move within a scene in two different ways—by using the teleportation and "move with arrow" option.

- **Teleportation:** On activating the Teleportation button, a teleport anchor will appear on the left controller. By pointing the anchor in any location and pressing the Grip button on the controller, the user can teleport to that specific location.
- **Moving with directional arrows:** On activating the Move with arrow button, an arrow prefab will be visible on the screen upon clicking the left controller Trigger button. While holding down the Trigger button to a specific threshold, the user will start moving in the forward direction of the arrow prefab. The user can also use the touch pad axis for continuous turn.

Marking By clicking the Marking button, the user can mark a specific location within the scene. To mark a place, holds down the Trigger button on the right-hand controller and moves around to create any shape.

Measurement tools To measure a distance between two points in a world space, select the blue arrow with a raycast and place it by holding down the Grip button on your controller. Adjusting the angle and scaling can be done using slider control. The distance value will keep updating depending on the arrow's position.

Placing 3D objects Clicking the Item button will open an object tray where you can select different objects. Clicking the Image button will make the object appear in the scene. To place the object, point the raycast at an object and drag it by holding down the Grip button. To remove the object from the scene, click the Remove button.

Labeling objects To label an object, the virtual keyboard needs to be activated, which will show text as you type keys with a typing stick. Select the add tag button to attach the text automatically on the object.

6.8 The AR UI Application

Plane Tracking To track a plane on a mobile device, we point the camera at any surface or floor. If the custom plane prefab, like the dotted plane in our scene, appears on the screen, it means the plane tracking was successful. The Toggle Horizontal Plane is turned on when it detects horizontal surface, Toggle Vertical Plane when it detects the vertical surface and a big plane when the dimension of a plane is greater than a specific value.

Image Tracking Every time an image from the reference image library is found, the prefab will be instantiated. The reference image library can be changed at runtime, but it can't be non-null as long as the monitored image management component is turned on.

6.9 Future scope and improvement

Working on a project implying Virtual Reality and 3D application can be a time-consuming and complex process, because developers have to deal with the physics, geometry, camera point of view etc. of 3D elements, etc. The system has a wide scope for improvement, as the current system is being sketched down to few modules. More features can be added such as eye tracking, motion tracking, infrared tracking, voice control, gesture control navigation, remote network multi user interaction, etc. in the future which may help the user to accomplish a larger number of tasks with more immersive experience and bring it closer to reality.

Furthermore, the current Augmented Reality environment uses basic features of AR like plane tracking and image tracking, making it more realistic and portable for mobile devices. More advanced features can be added later on to improve the system.

Finally, the system can evolve into mixed reality, which combines the real and virtual worlds to create a new environment and visualization in where real-world and virtual items can coexist and interact in real-time. [23].

6.10 Conclusion

In this study, we propose a comprehensive method for classifying and continually visualizing point clouds containing millions of points in a Virtual Reality (VR) and Augmented Reality (AR) environment. Additionally, we successfully imported a large .obj file containing a 3D object's coordinates, texture maps, polygonal faces, and other data with an import speed of around 750k triangles in about 10 seconds during runtime. We carefully handled user movement by implementing fine movement techniques, allowing for smooth movement without any loss of frame rate or slowness. The directional arrow movement provided freedom of movement within the world, while the teleport system enabled faster movement.

The marking and measuring system demonstrated excellent interactivity within the scene, with approximate calculations. The instantiation of several objects at once to create a paint brush effect was a significant achievement, and the distance measurement using interpolation between two points proved effective. While performance was a challenge due to the nature of our 3D application, the system could run on any computer with decent hardware specs. Nonetheless, we attempted to improve performance by clipping the camera view of our scene, but in future, we plan to introduce advanced concepts like Object Pooling to greatly enhance performance.

The open source Unity application we implemented displayed excellent performance in presenting enormous point cloud data. Our system offers a variety of tools and techniques, such as interaction, locomotion, and user interfaces, creating an immersive experience with the final product.

Bibliography

[1] R. T. Azuma, "A Survey of Augmented Reality," Presence: Teleoperators & Virtual Environments, vol. 6, no. 4, pp. 355–385, 1997.
[2] J. Carmigniani and B. Furht, "Augmented Reality: an Overview," In Handbook of Augmented Reality, pp. 3–46, 2011.
[3] M. Billinghurst, A. Clark, and G. Lee, "A Survey of Augmented Reality," Foundations and Trends® in Human–Computer Interaction, vol. 8, no. 2–3, pp. 73–272, 2015.
[4] S. Beutel and S. Henkel, "In Situ Sensor Techniques in Modern Bioprocess Monitoring," Applied Microbiology and Biotechnology, vol. 91, no. 6, pp. 1493–1505, 2011.
[5] M. Mekni and A. Lemieux, "Augmented Reality: Applications, Challenges and Future Trends," Applied Computational Science, vol. 20, pp. 205–214, 2014.
[6] A. Kharroubi, R. Hajji, R. Billen, and F. Poux, "Classification and Integration of Massive 3d Points Clouds in a Virtual Reality (VR) Environment," The International Archives of the Photogrammetry, Remote Sensing and Spatial Information Sciences, vol. 42, no. W17, 2019.
[7] Y. A. G. V. Boas, "Overview of Virtual Reality Technologies," In Interactive Multimedia Conference, vol. 2013, August 2013.
[8] P. R. Desai, P. N. Desai, K. D. Ajmera, and K. Mehta, A Review Paper on Oculus Rift-A Virtual Reality Headset, 2014, arXiv preprint arXiv:1408.1173.
[9] P. Tranton, "Samsung Gear VR: An Easy Guide for Beginners (Vol. 1)," Conceptual Kings, 2016.
[10] M. Borges, A. Symington, B. Coltin, T. Smith, and R. Ventura, "HTC Vive: Analysis and Accuracy Improvement," In 2018 IEEE/RSJ International Conference on Intelligent Robots and Systems (IROS), pp. 2610–2615, IEEE, October 2018.
[11] S. Keene, "Google Daydream VR Cookbook: Building Games and Apps with Google Daydream and Unity," Addison-Wesley Professional, 2018.
[12] D. MacIsaac, editor. "Google Cardboard: A Virtual Reality Headset for $10?" The Physics Teacher, vol. 53, no. 2, pp. 125, 2015.
[13] J. M. Zheng, K. W. Chan, and I. Gibson, "Virtual Reality," IEEE Potentials, vol. 17, no. 2, pp. 20–23, 1998.
[14] X. F. Hana, J. S. Jin, J. Xie, M. J. Wang, and W. Jiang, A Comprehensive Review of 3D Point Cloud Descriptors, 2018, arXiv preprint arXiv:1802.02297, 2.
[15] E. M. Mikhail, J. S. Bethel, and J. C. McGlone, "Introduction to Modern Photogrammetry," John Wiley & Sons, 2001.
[16] D. Girardeau-Montaut, CloudCompare. France: EDF R&D Telecom ParisTech, 11, 2016.
[17] J. K. Haas, A History of the Unity Game Engine, Diss. Worcester Polytechnic Institute, 483, 484, 2014.
[18] C. Coutinho, "Setting Up Your Project for VR Development," In Unity® Virtual Reality Development with VRTK4, pp. 13–25. Apress, Berkeley, CA, 2022.
[19] A. Gomes, L. Figueiredo, W. Correia, V. Teichrieb, J. Quintino, F. Q. da Silva, A. Santos, and H. Pinho, "Extended by Design: A Toolkit for Creation of XR Experiences," In 2020 IEEE International Symposium on Mixed and Augmented Reality Adjunct (ISMAR-Adjunct), pp. 57–62, IEEE, November 2020.
[20] W. Piekarski and B. H. Thomas, "Augmented Reality Working Planes: A Foundation for Action and Construction at a Distance," In Third IEEE and ACM International Symposium on Mixed and Augmented Reality, pp. 162–171, IEEE, November 2004.
[21] A. Sarkar, K. A. Patel, R. G. Ram, and G. K. Capoor, "Gesture Control of Drone Using a Motion Controller," In 2016 International Conference on Industrial Informatics and Computer Systems (CIICS), pp. 1–5, IEEE, March 2016.
[22] B. Guenter, T. B. Knoblock, and E. Ruf, "Specializing Shaders," In Proceedings of the 22nd Annual Conference on Computer Graphics and Interactive Techniques, pp. 343–350, September 1995.
[23] E. Bal, "The Future of Augmented Reality and an Overview on the to Researches: a Study of Content Analysis," Quality and Quantity, vol. 52, no. 6, pp. 2785–2793, 2018.

Lipsa Das, Vimal Bibhu, Ajay Rana, Khushi Dadhich, and Bhuvi Sharma

7 Convergence of AR & VR with IoT

Abstract: Augmented and virtual reality are two technological domains that provide and improve the real world simulation of the real objects. Internet of Things is another technological aspects which enables the integration of the physical objects upon the digital platforms. In this chapter we present the IoT based applications having the AR and VR technologies integration to enhance the sensitivity and quality of service of the applications in real world environments. The details of the devices based on the IoT are elaborated in context of AR and VR to create the mixed reality. Architectural frameworks of VR is defined with respect to market requirements like applications in the field of medical, entertainment, automotive, aerospace, and also for the electronic commerce. The useability and scope of the augmented reality with IoT for applications like connected vehicles, vehicle congestion management and future information of the congestion, smart grids, environment management, management of smart buildings and homes, smart cities, supply chain management and industrial, agricultural, and commercial applications and scopes are detailed for the future endeavour for the merged technologies of AR, VR and IoT. The technological frameworks of the IoT based objects and scalability of the AR and VR along with the IoT is detailed to focus on the communication and collaboration enhancements in industries and other public sector firms. Also, we present the challenges and cost pertaining to the IoT based Object simulation over the AR and VR technologies.

7.1 Introduction

The evolution of IoT-based technologies has made it possible to link multiple devices together using an variety of communications rules. IoT is a network of physical devices that are autonomously connected to each other, the internet, and users. One common use case for IoT is the remote monitoring of physical systems, which can be accomplished using sensors, surveillance cameras, and actuators. In the future, IoT devices are expected to be an crucial part of technology in various fields such as healthcare, transportation, etc. One of the most promising areas of IoT is in the smart building and the smart city industries, where it enables services such as energy management, building automation, and security.

Both service suppliers and users must be able to interact with IoT applications in a context-aware way, where they perceive and respond according to their surround-

Lipsa Das, Vimal Bibhu, Ajay Rana, Khushi Dadhich, Bhuvi Sharma, Amity University, Greater Noida, UP, India, e-mails: lipsaentc9@gmail.com, drvimal@gn.amity.edu, ajay_rana@amity.edu, khushidadhich08@gmail.com, bhuvisharma86@gmail.com

https://doi.org/10.1515/9783110785234-007

ings and user context. Nowadays, IoT is a general term that is used with various other terms such as Augmented Reality and Virtual Reality [1]. AR and VR are computer-generated simulations that create a 3D artificial environment around the user. In 2016 many technology-based companies began launching their AR and VR based systems, resulting in rapid growth. Augmented Reality and Virtual Reality are technologies based on reality that improves the real-life experience through simulation. Although their aim is the same, their applications are totally different. AR adds digital elements to the surroundings using a camera or smartphone, while VR is a mesmerizing experience which completely replaces the real environment with a parallel one [2].

The combination of IoT with VR and AR is a revolution that aims to integrate the physical and digital world, providing physical features to digital objects and also interacting with digital objects in the similar way that humans interact in the real world. It provides various organizations with opportunities and benefits, allowing them to enter this field. The increased use of IoT technology has resulted in a decrease in product ideation and delivery time. This convergence of AR and VR with IoT resulted in a new concept of Mixed Reality and thus a new for enterprises to work more towards future technological innovations. AR, VR and IoT are key technologies that enhance the effectiveness and efficiency of various fields and have the capability to renovate digital infrastructure. The future can be predicted through the execution of these technologies, potentially changing the present digital world [3].

Both the service suppliers and users must able to interact with IoT applications in a context-aware way, perceiving and responding according to their surroundings and user context. AR and VR offer an intuitive interface to IoT, enabling context, and overlaying virtual information in the real world, as with cognition-based architectures. It enables consumers to communicate with the amenities provided in a virtual way. AR and VR can be used to debug or repair the IoT devices which are not working properly by showing QoS-based measurements of the facilities provided by the gadgets, such as throughput and response time. Metrics of devices like the device temperature, CPU, and memory utilization can be displayed too. Intrinsic parts of gadgets can be linked to objects, allowing field workers to view specific visualizations for fixing malfunctioning internal components of an IoT-based device that is not working properly. This enables simple repairs and return to proper operational characteristics.

IoT-based applications may be used across various domains, which varying QoS requirements per domain depending on the application's sensitivities and importance. QoS can be classified into different types, such as best effort (no-QoS), differentiated services (soft-QoS), and guaranteed services (hard-QoS). In the case of guaranteed services, a hard, strict QoS is guaranteed, and it is used in security applications such as monitoring patients at the hospital or in a driverless car system. Whereas differential services do not ensure hard-real-time QoS, they are capable of fixing services that are failing. This can be a routing app that uses prediction of air quality, floods, and pedestrian movement to suggest optimal routes across a city. If any one of these services are on the verge of failure, the application would require the use of appropriate alternative services to be rebuilt. This scenario is ideal and does not guarantee the failure of any services. For example, atomic services can automatically monitor the temperature inside the home. The faults in any IoT services can be fixed by using an LSTM network [1].

IoT is expected to experience tremendous growth in the near future, with VR and AR likely to follow. VR and AR applications are also being implemented in the education sector; experts suggest that these apps can be efficiently used in field studies as well as in classroom learning. IoT, VR, and AR share a similar core idea: combining physical and digital elements to provide new experiences for consumers. While the Internet of Things digitally manipulates real-world objects, VR and AR render digital worlds to appear real. While IoT has a larger market, VR and AR are still at their infancy stages, but they are attracting many investors. Before we reach any conclusion, it is safe to say that IoT, AR, and VR could be converging on different fronts. Since these technologies share their core vision, they may be integrated to serve a variety of beneficial purposes.

7.2 IoT devices

An IoT device is any appliance which is connected to a sensor, is capable of collecting or sharing information with other devices and of performing tasks via the internet without human interaction. Many IoT objects can be controlled according to user needs, while others can control themselves based on specific situations.

There are various types of Internets of Things devices available in the market. Let's discuss some of the commonly used IoT devices.

7.2.1 Smart home appliances

These home appliances are designed to meet user requirements and assist them in their daily tasks, making them easier and more efficient. Examples of such devices are smart speakers, refrigerators, and many more. The smart home devices market reached $23.328 billion in 2020, making it the top contributor to the IOT devices market.

7.2.2 Industrial sensors

These devices are designed for industrial purposes, enabling manufacturers to gather insights about their machines or monitor issues at a factory using internet-connected sensor devices. IoT sensors can improve machine maintenance and operation visibility, as well as the utilization rates for specific resources.

7.2.3 Smart automobiles

The IoT capability is a critical component in building autonomous automobiles, cars and trucks of various sizes are using this technology. The investments in the self-driving car industry has exceeded $100 billion, and many companies are working on developing autonomous vehicles.

7.2.4 Smart cameras

Smart cameras are widely utilized for both personal and commercial purposes. Nowadays, many cameras are connected to internet to ensure that the recordings are securely stored on a local server or cloud.

7.2.5 Manufacturing robots

Robotics production is increasing rapidly, as robots are capable of performing various tasks more accurately. They can be remotely controlled and programmed to meet specific needs of manufacturers or users.

7.2.6 Wellness gadgets

As people becoming increasingly busy, they may not have the time to focus on their health. IoT technologies have developed wellness gadgets like fitness bands that allow users to monitor their health status. These devices collect data on various health metrics, making it easier to track progress and make improvements to lead healthier lives. [4]

7.3 IoT integration with AR and VR

The prevalence of IoT, AR and VR is unstoppable. Each technology provides many advantages to users either individually or in combination. Moreover, researchers are currently focusing on the future of these technologies. The impact of IoT's integration with AR and VR is nothing short of a revolution. It aims to combining the physical world with the digital one, not only to see things in reality but also to provide an environment where physical characteristics are provided to digital object. Essentially, it means the real grounding of digital objects into the physical environment and interacting with them in the same way as with the physical one.

Since IoT is trying to find new ways to interact with our environment, it provides multiple advantages to companies so that they can completely transform their business methods and revenue. We have already noticed a significant reduction in product ideation and delivery time with the help of IoT. The combination of these three different technologies results into the formation of a new concept or technology known as Mixed Reality (MR). MR is a technology that breaks all the barriers and limitations of the current technology and helps to create new possibilities. MR allows collaboration with companies on a global level, regardless of their location, opening the way for more technological innovations.

MR provides an environment for the user where the physical reality and digital objects are integrated in such a way that they provide interaction with the real world and virtual objects. Unlike VR, which immerses the user in a completely digital or virtual environment, or AR, which puts digital objects on top of a physical environment, MR blends the digital and real worlds into each other. MR is also known as Hybrid Reality or Extended Reality. It is sometimes referred to as AR, but its capacity for interaction between the physical world and digital objects places it further, with physical reality on one side and immersive virtual reality on the other.

The impact of IoT is on the entire environment. The same can be said for both AR and VR, which add an exciting and unique touch to our lives. They help to improve the effectiveness and efficiency of various areas and have the ability to completely renovate the automated structure. Nowadays, it is more exhilarating to see companies combining and merging these technologies and improving their services.

7.4 Architecture of VR systems

The formation of VR system is designed to support virtual environments, requiring expertise in diverse areas, from networks to psychology. Creating VR systems is costly in terms of time, human resources, and finances. They can provide a wide range of applications, such as training, scientific visualization, gaming, etc. This vast array of applications creates a set of requirements that make it challenging to develop a single system that fulfills all the requirements. As a result, monolithic systems have emerged, which are highly optimized for specific applications, but offer limited reusability of components for other purposes [5]

Several crucial components comprise a virtual reality system:

7.4.1 Viewing system

For the best virtual reality experience, a high-quality viewing system is essential. The viewing system connects users to the VR experience, irrespective of the number of users involved.

7.4.2 Tracking system

A sensor camera is required in VR headset to recognize movement and can provide an optimal 3D experience for the user.

7.4.3 Interactivity elements

A key aspect of VR is the ability for users to interact with digital objects as if they exist in the real world. In the early days, VR applications struggled to create a realistic environment, but the technology has since advanced and improved significantly. The interaction with digital elements depends on speed, mapping and range. VR can offer interactive features such as moving elements from one location to another and the ability to change an object's surroundings.

7.4.4 Artistic inclination

VR creates an environment that fully engages the user. When designing an environment that makes users feel like they are part of the virtual world, the designer should focus on entertainment value, engagement, and atmosphere.

7.4.5 Sensory management system

User must be able to perceive subtle changes occurring in the environment, such as movement, direction and vibration [6].

7.5 VR software architecture

The overall architecture of the proposed cost-effective VR systems follows a distributed software model, comprising one control node, called producer, as depicted in Figure 7.1, and one or more consumer nodes. The producer provides the GUI, and allows for the

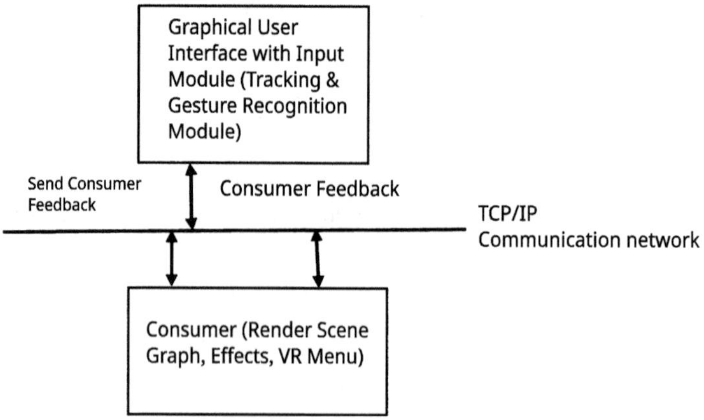

Figure 7.1: Proposed software architecture of the virtual reality system [7].

integration of additional input modules, such as those for user input via the tracking system and gesture recognition. Generally, the producer generates the user-input-based information to present to the user through the consumer.

The consumers perform actual scene-graph traversal and rendering, including the presentation of effects that describe the implied functions mapped to the virtual objects. They are also responsible for updating the scene graph according to changes resulting from user interactions.

The consumer showcases the VR Menu, which users can enable to execute various actions or specify how the system should interpret particular user actions. The producer-consumer communication is facilitated via TCP/IP. The producer maintains connections to each consumer, sending packets that inform the consumers about the user's specific commands.

The system is divided into a series of modules responsible for particular tasks, organized into packages, as illustrated in Figure 7.2. The producer app is responsible for receiving user input., providing a GUI for actions like loading the scene, changing the current viewpoint, or playing animations. The producer accesses the DTRackUDPReceiver module to receive user motion data from a tracking system.

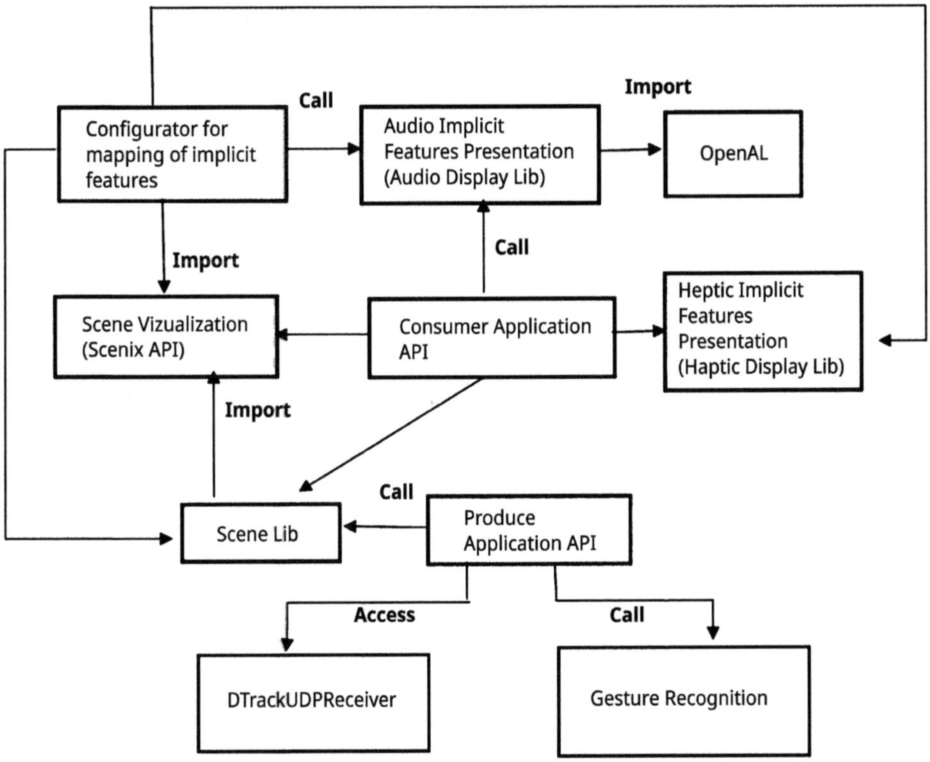

Figure 7.2: Decomposition of the proposed VR system architecture into packages [7].

It can then use this information to determine whether a gesture was performed by a user, calling the Gesture Recognition module.

When the producer receives user input, it sends the corresponding package to any connected consumer apps. The consumer applications render a virtual scene from the provided viewpoint, using the SceniX API to display the scene, and potentially employing the Audio Display Lib and HapticDisplayLib to present the implied features in audio and/or haptic formats. The configurator app maps the implicit features onto objects in a virtual scene, relying on the SceneLib library to load a basic scene diagram (without implicit features) which is shown to the user during mapping [8].

7.6 VR applications in IoT

The market is categorized into medical, entertainment, automotive & aerospace, e-commerce & retail, and other sectors based of its applications. The automotive and aerospace sector was the leading industry by size in 2015. In April, BMW Group (Germany) introduced mixed reality systems into car design, inspired by the computer game industry. After extensive R&D, BMW integrated the HTC Vive virtual reality headset and mixed reality technologies into their latest car models. With IoT, this is likely to result in semi- or full-automatic car-driving procedures.

We have discussed the application of VR in different sectors below:

7.6.1 Remote operations management

In remote locations where human contact is infeasible, such as adverse climate conditions, or outer spaces, crucial functioning can be managed more interactively by integrating AR in a smaller set-up environment. The exact series of activities occurring at a distant location can be visualized within a controlled conduit using AR, while live views can be employed for organizing and coordinating new tasks in real-time. IoT-enabled systems ensure that all the relevant environmental variables are assessed and evaluated for the feasibility of new ventures that may be launched.

7.6.2 Customer experience management

Nowadays, companies have various stages for prospect identification, networking, and conversion, leading to favorable trade prices. With IoT, lead nurturing personalization rates are enhanced. This is achieved through the ability to create insights and data from connected devices throughout a customer's physical location, such as smart wearables, and in-home assistants. IoT allows devices to route this information through large stores, enabling customers to find products they like quickly with an interactive search catalog.

They can even try products using IoT brightly-lit mirrors and directed vision, initiated by motion-controlled interactions with suggestions, products, etc.

7.6.3 Smart maintenance

One of the more critical areas where IoT finds functional relevance is in preemptive maintenance routines, such as those for hardware equipment. These range from factories, power and utility production sites, power grids, to smart home devices. Smart sensors may transmit information that prompts engineers regarding plant health, allowing for simple and straightforward tracking of maintenance tasks in large-scale server deployments. With the inclusion of VR expertise, technicians and support personnel can be guided to the precise component locations and receive interactive demonstrations of operations processes. These innovations enabled them to perform service tasks on the spot, without the constraints faced by remote teams working with specialists through video or audio communication channels. By completing maintenance tasks quickly and with expert guidance, companies find it much easier to control operating costs in the long run. In contrast, improper care schedules and ineffective maintenance processes performed without experts can result in losses on an accounting ledger [9].

7.7 AR applications in IoT

IoT connectivity is spreading globally across industries, as well as into individual vehicles and homes, where its implementation is most visible.

IoT applications with AR technologies are listed below:

7.7.1 Connected vehicles

Self-driving cars are among the most prominent examples of IoT in action. Autonomous vehicles utilize an array of connected devices to safely navigate roadways, regardless of congestion and weather conditions. Technologies used include AI-enabled cameras, motion sensors, and onboard computers. IoT connectivity is also present in traditional vehicles, where manufacturers install connected devices for monitoring operations and managing the computer systems. Municipal buses and commercial fleets such as delivery trucks are frequently equipped with more IoT-based technologies, such as interconnected systems for monitoring safety issues. Privately-owned cars and trucks may have similar technologies installed, typically provided by insurance companies to gather and relay telemetry data to verify proper driving habits. The Internet of Things has the potential to bring economic value to various sectors, including manufacturing, construction, transportation, and retail.

7.7.2 Congestion management

Over the past decade, highway infrastructure has become increasingly interconnected, with traffic light controls, cameras, sensors, parking meters, and mobiles-based traffic applications all relaying information that is then used to prevent congestion, avoid accidents, and ensure a seamless driving experience. Cameras, for instance, identify and relay data on traffic volumes to central management groups, who analyze the data to determine which mitigation measures should be taken. Traffic sensors can identify different levels of light in the sky and adjust signal intensity, helping to ensure visibility for drivers. Linked devices are used to identify available parking spaces and relay this information to a kiosk or an app for motorist notifications.

Monitors installed on bridges gather and relay data to analyze their structural health, alerting authorities of maintenance needs before any issues or breakdowns occur.

7.7.3 Smart grids

Utilities also leverage Internet of Things to improve the efficiency and resilience of their power grids. Historically, power flowed in a single direction on a grid: from generation sites to customers. However, connected devices now enable bidirectional communication across the entire power supply chain – from generation, to distribution, to usage – thus enhancing a utility's ability to move and control power.

Utilities can capture and analyze live information transmitted from connected devices to identify power outages, redirect distribution, and respond to fluctuations in power demand and load. Currently, smart meters in residences and offices provide insights into both real-time usage and historical usage patterns, which utilities and costumers can examine to identify strategies to increase efficiency.

7.7.4 Monitoring the environment

Connected devices can gather IoT data related to water, soil health and quality, air, fisheries, forests, and other natural areas, as well as weather, and other environmental. As a result, the Internet of Things not only provides access to a vast amount of real-time environmental data at any given location but also allows various corporations across multiple industries to leverage this information for actionable insights. These insights could help public agencies better manage and predict natural disasters like tornadoes, and to protect land and wildlife populations more effectively. Organizations can use this information to minimize their carbon footprint carbon, document environmental compliance or devise more efficient strategies for dealing with atmospheric conditions affecting their businesses.

7.7.5 Smart buildings and homes

Property owners are making use of the Internet of Thing's capabilities to make buildings smarter, meaning more energy-efficient, comfortable, cost-effective, and potentially healthier and more secure. IoT-based ecosystems in commercial buildings include HVAC management systems that use real-time data and automation technologies to continuously compute and adjust temperatures to maximize energy efficiency and comfort. Meanwhile, AI-based cameras can assist with crowd control to ensure people's safety during events like sold-out concerts. At home, individuals can install smart technologies like appliances, door locks, thermostats, and smoke alarms to help with daily needs, for instance, by coordinating temperature controls to a homeowner's schedule.

7.7.6 Smart cities

Smart cities integrate IoT deployments across multiple aspects, giving providing a unified view of all activities within their jurisdictions. Smart cities typically include interconnected transportation management systems as well as their technology-based buildings (i. e., smart buildings). They can also include privately-owned smart buildings and connect to IoT-based networks benefiting of environmental monitoring to build a large IoT ecosystem that provides a realistic view of various factors affecting resident's lives. Similar to smaller, more targeted IoT deployments, smart cities aim to collect real-time data for analytics, which provides insights that municipal officials can use to make better decisions or automate controls, leading to better-performing, efficient, resilient, and safer communities. For example, Copenhagen, Denmark, is using IoT technologies to achieve its goal of becoming a carbon-neutral city by 2025.

7.7.7 Supply chain management

Supply chain management systems have modernized thanks to low-power sensors, GPS and other tracking technologies that locate assets as they move through a supply chain. This data allows managers to plan efficiency and confidently inform stakeholders about items being dispatched or received. This kind of visibility is valuable, but it is only the beginning of the value proposition IoT brings to the discipline. IoT based technologies can monitor and manage shipment requirements, such as measuring the regulating temperature during transportation to ensure quality and safety controls. Additionally, back-end analytics capabilities can leverage the data obtained by the Internet of Things to identify improvements in the supply chain, like more efficient routes or of delivery times.

7.7.8 Industrial, agricultural, and commercial management

The Internet of Things has numerous benefits and uses in the industrial and commercial sectors, enabling everything from predictive maintenance to enhanced security and smart agriculture. These diverse use cases employ a wide range of Internet of Things technologies.

A manufacturer may utilize machine-to-machine interconnected devices within an IoT deployment for the industry to better manage workloads. A plant can monitor wear on machinery to schedule preventive maintenance at the optimal time. Organizations can manage and control access to their facilities using badges or RFID-enabled wearable devices. Farmers can integrate location-based technologies with environmental monitoring devices and their on-farm equipment to automate and optimize seed distribution.

7.8 AR enabled IoT platforms for a smart and interactive environment

Currently, Augmented Reality (AR) and IoT have attracted the attention of various researchers for creating smarter, more realistic, and interactive environments [10]. AR is an enhanced version of the real world that is achieved by using visual digital components, sounds, and many other sensory stimuli (hearing, smelling, touching, etc.) delivered through technology. Essentially, AR helps users understand certain aspects of the real world and brings smart, intelligent, and accessible innovations that can be used in reality. IoT is a vast network of connected devices (such as sensors, lights, thermostats, etc.) that can collect, share or exchange real-time data with embedded minimal computing elements for communicating or sensing [11].

AR and IoT can be considered complementary to each other. Firstly, AR provides an easy and user friendly way for users to visualize and interact with IoT-based devices and the information they contain. It offers a visual interface that is operable in the real world, comprehensible, and highly useful anywhere, anytime for users [10]. An AR user wearing a headset can quickly connect to an IoT device, receive object-related data, control data and associated AR data files for targeted services, understand how to interact with the current data from the IoT device, and finally, interact with physical appliances using direct controls by interacting naturally [14]. IoT is an infrastructure for AR that offers a systematic approach for AR to be "scalable" by handling data management (for instance, tracing data and content) in a properly distributed manner. Therefore, any IoT device can be locally tracked seamlessly, and also the scalable interface allows for location-based geographic and AR services. Augmented Reality users can connect directly to any IoT devices instantly anywhere, anytime through network connections, and after receiving context-related data sets. Relevant services (such as product details, ap-

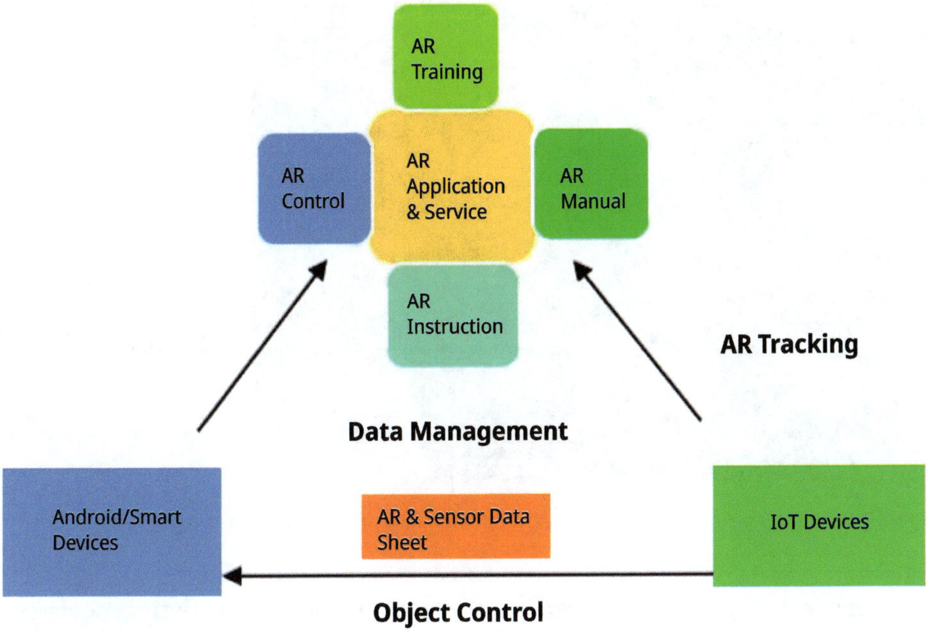

Figure 7.3: AR with IoT Integrated architecture [10].

pliance control buttons) are provided appropriately through AR glasses or mobile GUI, depending on the context data as shown in Figure 7.3.

Figure 7.4 provides an insight into how AR and IoT are combined in real-time. Ubiquitous computers and augmented interactions together form the system. Ubiquitous computers are computing devices that are easily accessible or can appear anytime in front of users. On the other hand, controlling in an Augmented Reality (AR) environment is referred to as augmented interaction. AR-enabled IoT means running IoT devices with AR visualization [13].

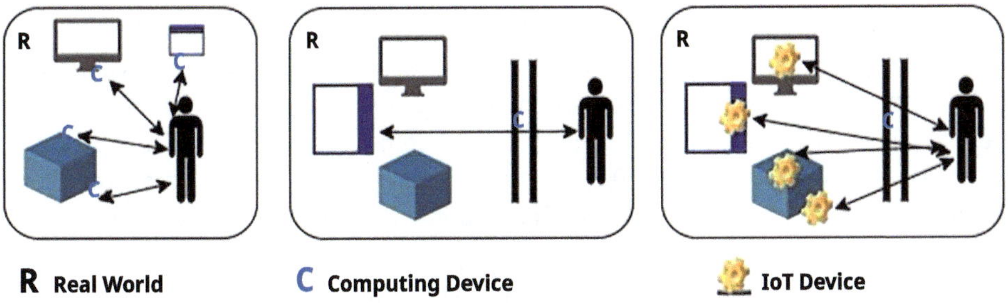

Figure 7.4: Demonstrating the combined system of AR and IoT with (a) ubiquitous computers (b) augmented interactions (c) AR-enabled IoT [13].

Figure 7.5: AR users wearing smart AR glasses interact intuitively (with hand gestures) [13].

For example, in Figure 7.5, a woman is working on her main project using AR glasses (computer-based glasses), which is displaying her all her work slides, videos, and information three-dimensionally by digitally generating them in the user's environment. Users can interact and retrieve information from mobiles, computers, and other devices digitally. The AR glasses also supports Wi-Fi, GPS, and Bluetooth.

7.9 Scalable AR recognition and tracking for IoT devices

In recent years, AR recognition has been an active area of research due to its potential and usefulness for context-aware computing. Augmented Reality (AR) studies focus on virtual 3-dimensional models. Recently, Augmented Reality technology associated with physical objects has greatly expanded. Intuitive visualization and features are closely related to AR technologies, allowing physical objects to easily interact with their surroundings. AR tracking methods involve analyzing relevant sensing data, with decisions based on the IoT devices. Claros et al. proposed an AR-based fiducial marker in the medical system to track real-time data combined with biometric signals of hospital patients. A Wireless Sensor Network (WSN) is used to process meaningful data gathered from distributed sensors, and the marker ID is overlaid with reality to visualize perceptual information.

Mihara et al. developed an LED Augmented Reality marker that reads LED blink patterns connected to a TV instead of fiducial markers. A significant issue with predetermined fiducial markers and blink pattern IDs is that there must be individual markers for each corresponding physical object. A large number of markers would be required for each physical object, if numerous objects surround an AR user, as well as extensive datasets for tracking. Therefore, an AR system featuring tracking will be more effective in enhancing the tracking performance due to the useful information associated with natural characteristics.

Secure recognition, speed, accuracy, and spatial tracking of objects are considered significant technical challenges for Augmented Reality (AR). A well-known approach is the feature-based method that recognizes an object and calculates its pose based on geometric features and their attributes detected by sensors. Various feature-based techniques may be applied to specific classes of objects and can also differ in their robustness.

Template image matching is often used with physical objects (without rich features like texture-less objects). However, using these methods in robust 3D tracking is not favorable. In situations with poor lighting, partial occlusions, and cluttered backgrounds, texture-less 3D objects are used for tracking, but these remain quite challenging. The model-based tracking method involves fitting and matching a 3D wireframe structure with edges extracted from a camera image to recognize and track targets. However, model-based approaches typically require complex solutions and lengthy intersection processes, and it is unclear which 3D wireframe model would be the most appropriate. Despite these complications, they are still feasible alternatives in the absence of feature information as shown in Figure 7.6.

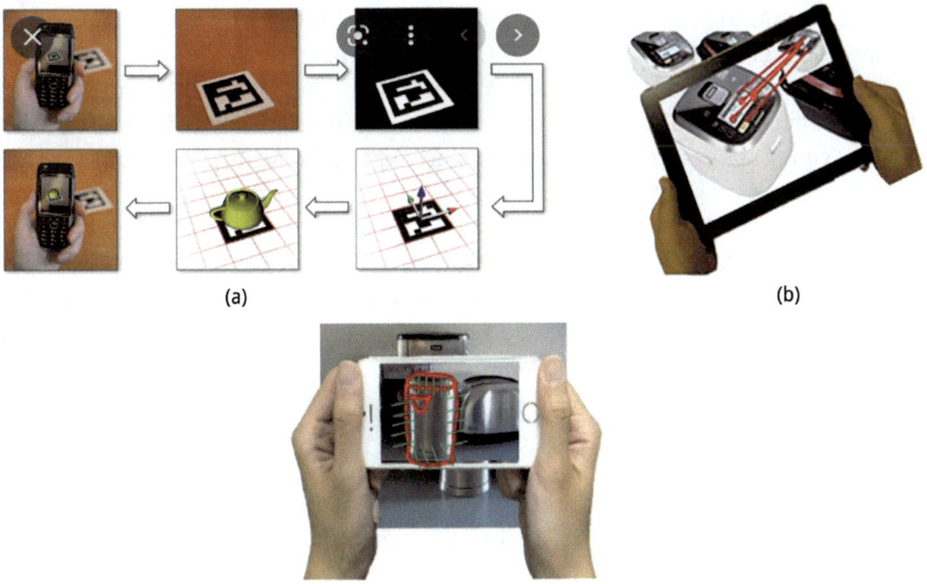

(a) (b) (c)

Figure 7.6: AR object for recognition and tracking (a) fiducial marker (b) feature based (c) model based [10].

7.10 IoT object control with scalable AR interaction

The widespread use of IoT includes object control [12]. Internet of Things devices consist of various sensors typically used for incoming data, actuators for controlling the object's functionality, and networking modules for wireless capability. Users can operate actuators through interaction, for instance, by turning on lights, fans, air conditioning etc. In a study carried out by a renowned researcher, he demonstrated how to control the behavior of objects (e. g., TV, refrigerator) using a simulated environment for home automation.

Recent studies show that sensors (or actuators) embedded in everyday settings provide visual representations of the connected modeling results produced. The dual-reality systems presented by Lifton and Paradiso produce interactions between a modeled world and an array of sensors, similar to electrical strips in the real world. A variety of experiences accompany this dual-reality simulation.

The focus is on mutually reflecting sensory navigation and interaction techniques with sensors connected via a sensor network. Bidirectional mapping techniques to enhance the visualization of IoT-enhanced information are proposed by Lu. Whenever the user activates the device within a real-world context, e. g., sensors function to sense user activities that are transferred into the virtual/simulated world. Then, virtual worlds can produce equivalent representations of it in real-world environments. This system is developed to achieve ecological feedback for energy conservation.

An excellent way of controlling objects with AR is using it to control both on-site and remote objects through cameras. Augmented Reality technologies can be used to display simulated control applications, such as for training purposes.

Recently, some attempts have been made to combine AR or VR as the interface to drive and simulate IoT objects. Additionally, there has been some experimentation with the direct control of everyday objects within an Augmented Reality environment. The real experience of users depends heavily on AR devices.

Many recent works have primarily relied on smartphones for images, merging the real and virtual worlds. Helmet-type head-mounted-display devices (HMDs) for augmented reality that overlay spatially registered virtual objects over the user's vision have been introduced. Helmet-mounted AR objects are generally classified as video see-through HMDs and optical, based on whether actual images are seen by the user or a video input. Video see-through HMDs attach two webcam modules to an HMD display, providing two sources of images: the real world and the virtual, digital world. Alternatively, an optical see-through HMD has the ability to blend virtual objects so the user can view them, i. e., a virtual world. Therefore, the focus is on the framework for interaction in AR, considering features of devices interacting with IoT devices, and the influence of different types of AR objects.

7.11 Other applications of AR and VR

AR and VR are making significant inroads into businesses, e-learning, healthcare, manufacturing, and various other fields. According to a Statista survey, Virtual and Augmented Reality will reach $296.9 billion in revenue by 2024 as shown in Figure 7.7.

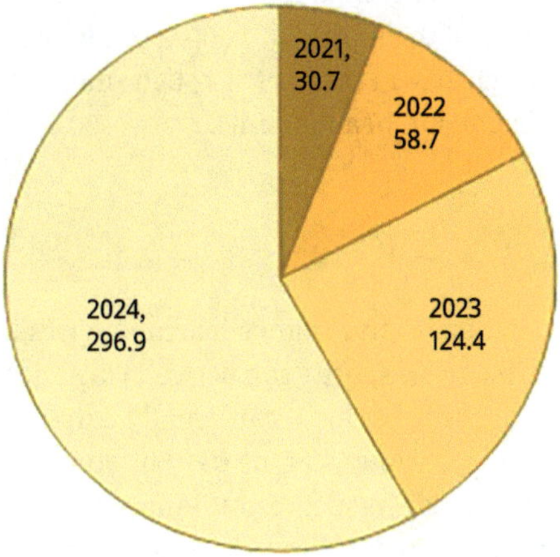

Figure 7.7: Market size of AR and VR worldwide (2021–2024) (sales in USD) [15].

Other important applications of Augmented and Virtual Reality are:

7.11.1 Enhancing communication and collaboration in industries

AR and VR play a major role in removing barriers in the international collaboration. They assist with translations and create shared spaces. People can easily attend meetings or live events virtually, particularly for those participants who cannot attend in person due to distance or other constraints.

7.11.2 Elevating user experience

Augmented Reality (AR) technology has made user experience more effective and enjoyable. People are showing great interest in online items like clothing, jewelry, housing, furniture, cars, etc. This has resulted in appreciable positive feedback and a significant increase in sales.

7.11.3 AR in the military

IVAS (Integrated Visual Augmentation System) was developed by Microsoft in collaboration with the US army to improve military forces' battlefield navigation, situational awareness and productivity.

IVAS is designed to help connect the battlefield with command centers, allowing soldiers to share critical information instantly and provide crucial situational awareness even when relying solely on their smartphones. One of the most vital tools a soldier can have in the field is night vision, which helps to see clearly in the dark through smoke or fog, thanks to multiple front-facing cameras embedded in head mounts.

7.11.4 Practical experiments

The lack of practical instruments is one of the challenges in distance learning. Virtual Reality and Augmented Reality have made it possible to bridge the gap between distance and real-life learning. Practicing virtually, like in-person learning, can greatly enhance students' understanding. VR headsets, along with headphones, can be used by students to perform experiment or tasks virtually and view models (such as buildings, organs, atoms or molecules) three-dimensionally, which can also develop their interest in the subject, providing them with a better learning environment.

This is not only useful for students but also for teachers, making it easy to demonstrate a chemistry experiment, for example. Pair-programming is one of the great concepts that allow the teacher to share their virtual screen and explain the procedure.

7.11.5 Overcoming language barriers

A very common problem that almost everyone has faced is the language issue. Many times, the person you are speaking with may not understand your language well, leading to confusion. This problem often arises in call centers due to poor skill-based routing, where the agent you speak with can't fulfill your request. Learning multiple languages can be a daunting task for anyone. Fortunately, Virtual Reality (VR) offers a solution: it can instantly translate and provide subtitles for the speaker, enabling people to learn and grasp the other person's words at a faster rate.

This is incredibly useful for international students during their online courses. Collaboration is another major aspect of AR and VR. Many foreign students often hesitate to speak as they fear mispronouncing and making grammatical mistakes. VR interactions help them reduce their mistakes and encourage many people from different backgrounds to socialize.

7.11.6 Augmented reality in search

Viewing 3D animals from home in real-time is fascinating. Google allows users to search and view life-sized animals three-dimensionally and interact with them in our local space using AR technology. It is supported by an ARCore-enabled devices. Snapchat is another great example of AR search. Users can see, move, and hear animals (bark, woof, meow, snort, etc.) and change the size of 3D animals like zebras, giraffes etc. by simply pointing their at the ground. AR technology also enables users to view cartoon characters and avatars.

7.12 Challenges in AR/VR applications

Augmented and Virtual Reality enhance our environment by adding digital components to a live user view. They provide an immersive experience that replaces the real world with a virtual one. While these technologies are evolving, there still are some challenges that need to be addressed. 99 % of people reported technical challenges that had to be overcome for AR/VR, according to the Augmented and Virtual Reality Trends Survey.

7.12.1 Latency

Latency while drawing new content is a major technical challenge for AR and VR. Every system has a threshold latency, which depends on three factors: the frame rate of the content being drawn, the display's refresh rate, and the input lag that occurs from the interaction. Problems arising from delays are not limited to AR and VR. Systems with a similar interface for interaction between humans and their respective content drawn on a monitor also face this problem. As AR/VR expands beyond the entertainment to play a significant role in healthcare, the military, etc. (e. g. doctors across the world are using these AR/VR technologies to deliver their medical services remotely), the latency problem must be resolved for their potential to be realized.

7.12.2 Cost and streaming

The biggest challenge of VR/AR is the high cost. VR devices like HTC Vive and Oculus Rift cost more than $500. Google Cardboard is a project that aims to provide users with a cost-effective experience, but at the expense of quality. Streaming is also a cost factor. Detailed content, high delivery rates, and the required bandwidth for streaming make it more expensive.

7.12.3 Compatibility with other hardware

All virtual headsets have unique content formats. Meaning the content format of Oculus Rift will not run on HTC Vive without modifications. A consolidation body is needed for AR/VR so creators can develop content for various platforms easily. It might be too early for such initiatives, given the VR's nascent stage, but it is certainly the way forward to ensure users have adequate choices. Augmented Reality is also moving forward on a similar path.

7.13 Conclusion & future directions

Since AR/VR is a cutting-edge technology, it has become challenging to represent AR devices for consumers. In the AR research field, all work or research projects are handled, and data is stored using predefined tracking information (for instance, feature sets by known fiducial markers) rather than loading the information online. The physical, which carries a web address along with it, is used to connect with smart devices and is considered one of the near futures of Augmented Reality.

This data exchange approach for IoT devices enables on-demand interaction with direct access to sensors related to everything in nearby areas. Sensors powered by high-speed networks, similar to AR-based environments, allow consumers within an Augmented Reality (AR) environment to get the natural AR experience instantly. They use a communication technique for the AR data sets essential for recognition and tracking anywhere at any time. To associate destination objects with AR information, AR systems require standardized data formats, such as function sets, augmenting content, and UIs. Therefore, the visualization and operation of AR within an IoT object app would need standard formatting structures of the attributes of 3D objects (e. g., functions, authorizations, behavioral animations, and filtering protocols). Moreover, the environment of Augmented Reality display and interaction with Internet of Things services (e. g., social interactions) is built upon the standard event processing procedures to achieve cross-platform interoperability with major portals like WebGL.

Additionally, IoT data sets may be too large to handle, when information is obtained over time. Thus, future studies are going explore effective AR visualizations methods for visual information in the context of real-time big data analytics (e. g., AR entity sizes, shapes, population densities). Then, AR technology supported by the Internet of Things will expand to development, focusing on filtering objects to consider context information of surrounding areas for users for the purposes of the AR services between objects and users. With this user-centric AR filtering approach, an AR consumer can easily connect with the scannable object.

Another challenge arises in complex situations where, to deal with numerous IoT objects in daily life, the vast number of Internet of Things based gadgets present in the near environment, AR Content is required to determine and optimize the representational complexity of a visual object associated to an IoT-based object according to its perspective-relative location. Optimized level of detail (LOD) techniques are useful for improving rendering efficiency while reducing the computing load of the AR-based computing device. Performance in a large-scale setting like IoT is reflected in the time required to find and map a destination entity and manage or evaluate interrelated information and content across many entities across a network. IoT-based objects with their own computation, networking, and storage abilities, content can stored, distribute and share content (without any requirement of internet infrastructure). Instead, IoT-based objects can automatically transmit required information to users automatically based on a need-to-know basis (including information needed for recognition and tracking). This means information is now assigned and distributed across various objects within the environment, resulting in faster and more reliable recognition and tracking of objects.

The AR client senses the presence of IoT devices (equipped with basic computing, storage, and networking components) within its immediate vicinity (similar to identifying Wi-Fi access points), and those objects transmit necessary location information back to the client. Because relatively few targeted objects are bound around, the client is quickly able to locate (and even track) an object and extract associated content. Note that there is no server or Internet involved in this implementation. One might suggest that IoT devices should be organized geographically and controlled via a hierarchical network of servers (similar to geographic services). However, considering the vast number of objects that must be managed (even when compared with geographical objects), no general-purpose technology currently exists for accurate tracking and recognition of individual objects (which could be mobile) at an indoor location. IoT creates an ideal framework of sensors to enable an "anywhere" experience of physical devices. By having the data stored locally on devices, the IoT ecosystem can quickly adapt to changing needs and requirements, paving the way for innovative solutions and revolutionizing the way we interact with technology in our daily lives.

Bibliography

[1] G. White, C. Cabrera, A. Palade, and S. Clarke, "Augmented Reality in IoT," In CIoTS 201, Hangzhou, China, https://doi.org/10.1007/978-3-030-17642-6_13.

[2] B. Srimathi, E. Janani, N. Shanthi, and P. Thirumoorthy, "Augmented Reality Based IoT Concept for Smart Environment," Int. J. Intellect. Adv. Res. Eng. Comput., vol. 5, pp. 809–812, 2019.

[3] E. Bastug, M. Bennis, M. Medard, and M. Debbah, "Toward Interconnected Virtual Reality: Opportunities, Challenges, and Enablers," IEEE Communications Magazine, vol. 55, no. 6, pp. 110–117, 2017, https://doi.org/10.1109/MCOM.2017.1601089.

[4] I. Lee and K. Lee, "The Internet of Things (IoT): Applications, Investments, and Challenges for Enterprises," Business Horizons, vol. 58, no. 4, pp. 431–440, Jul. 2015, https://doi.org/10.1016/j.bushor.2015.03.008.

[5] "Architecture of Virtual Reality Systems," In Stepping into Virtual Reality, pp. 107–116, Springer London, London, 2008, https://doi.org/10.1007/978-1-84800-117-6_5.
[6] https://www.izmofx.com/article/5-important-elements-of-virtual-reality-vr-2156-en-us.htm.
[7] S. Maleshkov and D. Chotrov, "Affordable Virtual Reality System Architecture for Representation of Implicit Object Properties," IJCSI International Journal of Computer Science Issues, vol. 9, pp. 23–29, Jul. 2012.
[8] SceniX, http://developer.nvidia.com/scenix-details. Last accesses 02.03.2012.
[9] M. Hu, X. Luo, J. Chen, Y. C. Lee, Y. Zhou, and D. Wu, "Virtual Reality: A Survey of Enabling Technologies and Its Applications in IoT," Journal of Network and Computer Applications, vol. 178, 102970, 2021.
[10] D. Jo and G. Kim. "AR Enabled IoT for a Smart and Interactive Environment: A Survey and Future Directions," Sensors, vol. 19, no. 19, p. 4330, Oct. 2019, https://doi.org/10.3390/s19194330.
[11] https://www.insiderintelligence.com/insights/internet-of-things-devices-examples.
[12] K. Michalakis, J. Aliprantis, and G. Caridakis, "Visualizing the Internet of Things: Naturalizing Human-Computer Interaction by Incorporating AR Features," IEEE Consumer Electronics Magazine, vol. 7, pp. 64–72, 2018, https://doi.org/10.1109/MCE.2018.2797638.
[13] The Physical Web. Available online: https://google.github.io/physical-web/ (accessed on 14 August 2019).
[14] https://www.analyticsinsight.net/ar-iot-together-enable-smart-interactive-environment/.
[15] https://www.statista.com/topics/3320/statista-surveys/.

Aditya Singh

8 Augmented Reality and its use in the field of civil engineering

Abstract: Augmented Reality is a technology that can create a real-world-like environment where all objects are generated with the assistance of a computer. The user can comfortably interact with these objects without disturbing the actual world in the slightest. In this book chapter, we focus on Augment Reality in a general sense and specifically in the context of civil engineering. The chapter not only explains the concept of Augmented Reality but also describes its brief history and evolution. Furthermore, it highlights the numerous advantages of using Augmented Reality, especially in the area of civil engineering, and discusses some major challenges. The author collected data from various sources to conduct a graphical analysis to support the study. Major technologies and examples related to Augmented Reality are also mentioned in this chapter. It will also support the main idea of the collection—a handbook on the interactive utilization of technologies like Augmented Reality or Virtual Reality in the real world.

8.1 Introduction

Augmented Reality (AR) is an interactive real-world experience where objects in the real world are enhanced by computer-generated perceptual information, sometimes through multiple sensory modalities, including visual, auditory, tactile, somatosensory, and olfactory. Augmented Reality can be defined as a system that includes three basic functionalities: a combination of real and virtual worlds, real-time interaction and accurate 3D registration of virtual and real objects. Overlapping sensory information can be constructive (i. e., enriching the natural environment) or destructive (i. e., masking the natural environment). This experience is closely related to the physical world, so much so that it is seen as an immersive aspect of the real environment. In this way, Augmented Reality alters the ongoing perception of the real environment, while Virtual Reality completely replaces the user's environment with a simulated one. Augmented Reality is associated with two widely understood terms: mixed reality and computer-mediated reality [36].

Acknowledgement: This research received no specific grant from any funding agency in the public, commercial, or not-for-profit sectors.

Aditya Singh, School of Civil Engineering, Lovely Professional University, Phagwara, India, e-mail: aditya7777777k@gmail.com, https://orcid.org/0000-0001-9347-5627

https://doi.org/10.1515/9783110785234-008

8.1.1 What is Augmented Reality in construction?

Nowadays, Augmented Reality (AR) is a frequent topic in discussions about the adoption of technology in the construction industry. In fact, AR connects the real and virtual worlds. Unlike its virtual reality counterpart, AR does not offer an immersive out-of-body experience. Instead, it typically involves the use of goggles, masks, or glasses to superimpose computer interfaces over the physical world. This flexible technology is rapidly gaining ground in the industry, as its applications range from collaboration to design to safety. Augmented Reality is used in the industry to increase productivity, improve safety, accelerate collaboration, manage costs and build trust in projects through the ability to deliver information in real-time. Here are some examples of how construction crews use Augmented Reality to speed up the construction process. AR combines the real and virtual worlds into one immersive experience. AR displays 3D images in the real environment of a person walking through space using a smartphone or an AR headset. AR technology can offer geographic information to users via GPS and cameras, providing relevant information as the user moves around the workplace. Schedules, operating data and drawings are readily available, allowing users to make selections on the fly as they work. Augmented Reality can affect stakeholders' perceptions of specific activities, but it can also change the way professionals work, saving time, energy and money [38, 39].

8.1.2 What are the features of Augmented Reality?

Augmented Reality presents a series of major features, which are discussed in the following:
- AR works according to the situational requirements.
- AR allows virtual world as well as physical world to amalgamate.
- AR utilizes all three dimensions
- AR allows the user to interact in real-time.

8.1.3 Objectives

This book chapter has several key objectives, which include:
- The book chapter briefly explains the concept of Augmented Reality in general, as well as in civil engineering in particular.
- This book chapter discusses the numerous benefits of using Augmented Reality in the area of civil engineering as well some major challenges of using it.
- The chapter will also highlight a thorough literature review of numerous scientific and research papers as well as the gaps in the research.

- This chapter also describes some major technologies which can be helpful in the field of civil engineering.
- It will also help readers understand how Augmented Reality can transform civil engineering.

8.2 Motivation

The author chose to study Augmented Reality because it is currently a growing trend, and with ongoing technological advancements, continuous improvements are being observed in the field of Augmented Reality. As a civil engineer, the author's primary focus is on applying technologies like Augmented Reality to the domain of civil engineering. Over the years, Augmented Reality has found increasing applications in civil engineering. This book chapter aims to expand the understanding of readers, including civil engineers, researchers, and the general public, about how Augmented Reality is not just limited to electronics, computer science, and gaming but can also be efficiently applied in other fields, such as civil engineering. As technology continues to advance, further improvements in Augmented Reality and are expected, which will enhance its applications and change the way civil engineering projects are carried out, including structural engineering, construction, and transportation projects.

8.3 History of Augmented Reality

In 1901, author L. Frank Baum first mentioned the concept of an electronic display or spectacles that could overlay information on the real world (as experienced by people). This was a Character Marker and also the beginning of the concept of Augmented Reality in recorded history. Between 1957 and 1962, Morton Heiling patented a simulator called Sensorama, which incorporated smell, visuals, vibrations, and sound. In 1968, Ivan Sutherland invented a head mounted display. By 1975, Myron Krueger created a Videoplace. Starting in 1980, significant work and developments began in this field. In 2015 Microsoft announced the HoloLens AR headset as well as Windows Holographic. The following year, Niantic released Pokemon Go for Android and iOS, which became a very popular AR game worldwide. In 2018, the Magic Leap One headset was announced by Magic Leap, which also used digital lightfield technology implanted in their headset. By 2019, Microsoft announced HoloLens 2, representing another major development in the history of Augmented Reality [36].

8.4 Literature review

Bellalouna (2022) investigated Augmented Reality focusing on the application of optical see-through AR technology in engineering contexts. She used HoloLens, an Augmented Reality device, in her study. She stated that the developed Augmented Reality apps were later implemented as prototypes in the Augmented Reality lab by the students of a German university [1].

Rohil and Ashok (2022) studied AR in the context of three-dimensional layout plans of an urban development. They analyzed two urban planning scenarios where new structures were created, and existing designs were replicated. They conducted their study in India and utilized BIM combined with AR to achieve the desired results [2].

Settimi et al. (2022) researched the drilling of timbers and smart retrofitted tools with the assistance of AR in Switzerland. The focus of their study lies on an initial assessment of potential bottlenecks, detailed technical challenges, fabrication systems assisted with tool-aware Augmented Reality as well as proposed object accuracy [3].

Behzadan et al. (2015) investigated AR visualization and its system application in civil infrastructure. Conducted in the USA, their study reviewed significant issues in Augmented Reality and explored technical approaches to address the major challenges hindering the practical implementation of this technology in CIS applications [4].

Alirezaei et al. (2022) studied risk management using a combined BIM-AR method. They aimed to enable an online assessment of schedule risks as well as project cost. Their proposed system could be implemented in real projects. They further claimed that 67 % of the survey participants acknowledged improvements resulting from its use [5].

Sidani et al. (2021) examined BIM-based AR and related techniques and tools. Conducting their study in Portugal, they performed a systematic review of past research to support their investigation. Their review methodology was based on PRISMA – P, and they selected 24 publications for their analysis, out of 671 publications like Web of Science, Scopus, etc. [6].

Delgado et al. (2020) studied AR as well as VR focusing on a research plan for construction, engineering, and architecture. They conducted several investigative workshops and surveys in the UK involving 54 professionals from 36 organizations, including both academia and industry, to obtain their desired study results [7].

Muthalif et al. (2022) researched AR and reviewed visualization methods for subsurface utilities. Conducted in Australia, their papers covered the current visualization methods of Augmented Reality, downsides of these methods, categorization, possible solutions, research gaps, and associated challenges [8].

Gomez-Jauregui et al. (2019) examined mobile AR applications and performed a quantitative assessment of occlusion inconsistencies in facility management, construction, engineering, and architecture. Conducted in Spain, they proposed a new methodology for mathematically executing quantitative assessments to obtain desired results [9].

Kwiatek et al. (2019) studied spatial cognition and AR, focusing on their impacts on assembly in the construction domain. Their research, conducted in Canada involved experiments which involved 40 engineering students and 21 professional pipe fitters. The cognitive abilities of the participants were measured. They were then required to assemble a complex pipe spool with the assistance of AR and also using the conventional procedure, in order to produce the projected study results [10].

Vargas et al. (2020) examined AR and its potential to address current challenges and future opportunities in the shipbuilding industry. Conducted in Norway, their study involved a thorough review of peer-reviewed empirical publications on Augmented Reality related to the shipbuilding industry, aiming to ensure proper utilization of Industry 4.0 technology [11].

Chi et al. (2022) investigated laser scanning and AR, working to integrate them for the examination of rebar. They conducted a lab-based experiment to confirm the success of their proposed method, suggesting that it could provide a practical solution for professionals to accurately and effectively inspect rebar in the industry [12].

Harikrishnan (2021) explored the feasibility of AR technology for communication purposes in the construction industry. Conducted in the USA, the study involved semi-structured interviews with current practitioners and a thematic analysis of the collected data to obtain the necessary results [13].

D'Anniballe et al. (2020) researched AR's potential use in air accident investigation and practitioner training. They argued that AR could effectively provide a full three-dimensional representation of an actual plane crash and facilitate training for the relevant professionals. Their paper emphasized the need for further research to overcome existing limitations and enhance the technology's capabilities [14].

AlFadalat and Al-Azhari (2022) studied AR in conjunction with architectural procedural modeling for the assimilation of contextual methods, specifically for residential buildings. Using Amman, Jordan as a case study, they proposed a general framework and tested their method on a group of university students. They also employed a machine learning model for evaluation [15].

Haemmerli et al. (2021) investigated AR and standard neuro-navigation, assessing their impact on the accuracy of perifocal structures during neurosurgical procedures. Their study, conducted in Switzerland, involved assigning tasks to 23 participants to compare performance and outcomes. The results showed that AR was highly favored over NV [16].

Hu et al. (2021) examined AR applications for teaching structural systems to non-engineering students. Conducted in Singapore, their study focused on AR's effectiveness through a quasi-experiment and a survey to obtain the desired outcomes [17].

Gazzotti et al. (2021) researched AR and VR applications in fusion design engineering. Conducted in France, their study utilized Computer-Aided Design modeling analysis and cinematic virtual reality simulations, along with the introduction of Augmented Reality [18].

Arulanand et al. (2020) investigated AR's potential to enhance the learning experience in engineering education. Their study, conducted in India, proposed a mobile-device-compatible AR framework and an experimental setup to transform a specific topic into an Android app to achieve their study objectives [19].

Wang et al. (2022) studied AR and its application in training, manual assembly, and repair instructions. They performed a comprehensive review of the current state of research, technical characteristics of manual operation instructions, past projects, and recent works. They claimed that their study would help researchers design Augmented Reality instructions in the future [20].

Sangiorgio et al. (2021) studied AR in decision-making to assist multi-criteria analysis in the construction field. They proposed a new Multi-Criteria Decision Analysis based on the hierarchical structure of the Analytical Hierarchy Process. They compared the traditional method with two enhanced versions of the AHP procedure to demonstrate the effectiveness of their proposed approach [21].

Xiang et al. (2021) researched cooperative robots and mobile projective AR for use in the construction industry. Conducted in the USA, they designed an algorithm and framework for their study. They reported that their assessment showed cm-level projection accuracy for mobile projective AR, highlighting its effectiveness [22].

Li et al. (2018) conducted a comprehensive review of AR and VR concerning safety aspects in the construction industry. They identified research gaps after examining numerous scientific and research publications, claiming that their results would benefit industry professionals and researchers in construction safety [23].

Kim et al. (2017) applied AR to wearable devices for developing a construction hazard avoidance system. Their study, conducted in South Korea, proposed an image-based system to alert workers to potential life-threatening situations at construction sites, aiming to enhance safety [24].

Pranoto et al. (2019) explored the use of AR to improve the learning experience in rail transportation. Conducted in Indonesia, they focused on train technology subjects during their experiment, claiming that their results could facilitate the practical implementation of AR in teaching and learning, especially for complex subjects [25].

Diez et al. (2021) studied AI, AR, and Big Data to assess transportation events. Conducted in Spain, they analyzed different technologies applied to create data verification and information related to transportation [26].

Yount et al. (2022) investigated AR for route learning and navigational assistance. In their US-based study, they created driving situation simulations to determine how route learning and driving performance of 62 adult drivers could be affected by the type of device used for navigational assistance. Their findings suggested the effectiveness of their approach in real-life situations [27].

Lovreglio and Kinateder (2020) researched AR for evacuating pedestrians in emergencies. They reviewed relevant publications to help people enhance the evacuation process from buildings during disasters such as tsunamis, earthquakes, or fires [28].

Bagassi et al. (2020) studied AR interfaces and conducted a human-in-the-loop assessment in the context of an airport control tower. They claimed that their proposed method would be particularly helpful when operating under low visibility conditions [29].

Smith et al. (2021) investigated AR's graphic spatial location and motion to determine their impact on driver behavior. They used four different graphics and 22 participants in their experiment, asking them to navigate a simulated environment using the provided graphics. They found that fixed graphics were better than animated ones in visually complex urban environments [30].

8.5 Research gaps

The comprehensive review of various scientific and research papers published in the recent years revealed that extensive research has been conducted on Augmented Reality and its applications in the area of civil engineering, but up to a certain extent. There is noticeable research gap in use of Augmented Reality in various sub-branches of civil engineering, like geotechnical engineering, water resource engineering, hydrology or wastewater engineering. There is still a considerable need of research work in traffic engineering or transportation engineering. Some major research work has been noticed recently in the applications of Augmented Reality in construction projects, but it is barely enough. There is also a lack of research in the design structures, as part of structure engineering, a sub-branch of civil engineering.

8.6 Main focus of the chapter

The author of the chapter focused on Augmented Reality, its applications, and challenges in the civil engineering field. Some significant examples and technologies of Augmented Reality are also mentioned in this book chapter. A comprehensive review of numerous scientific and research papers published in the recent years was conducted by the author to find the gaps in ongoing research. The author also collected data from numerous sources to perform a graphical analysis to support the study. Additionally, a systematic methodology was created as he highlighted the major benefits of using Augmented Reality in the area of civil engineering, which will be helpful in real projects including construction and transport projects. Overall, the work of the author aligns with the main idea of the book, which is to serve as a handbook for interactive utilization of technologies like Augmented Reality or Virtual Reality in the real world.

8.7 Issues, controversies, problems

In this section, the author discusses some major issues and problems related to use of Augmented Reality in civil engineering. While Augmented Reality has the potential to transform the field, there are still some improvements needed. One issue is that devices compatible with Augmented Reality are sensitive to weather changes, and bad weather can adversely affect their performance on site. Additionally, a good internet connection is a must in using AR, but in many civil engineering projects, people have limited or no access to decent internet connection due to insufficient infrastructure. Civil engineers often work in terrains that do not support internet connection, making Augmented Reality technology unusable on site. Another issue is the cost of using Augmented Reality, which is definitely higher than the conventional ways of working on a civil engineering project. As a result, many contractors, subcontractors, and engineers aiming to reduce the costs of the project will be hesitant to use Augmented Reality. Additionally, people require sufficient skills to use Augmented Reality comfortably, and compatible devices are needed at the site when working on projects, whether it is a transport, construction project etc.

8.8 Popular Augmented Reality technologies

In this section, some notable Augmented Reality technologies will be discussed that are specifically used in the field of civil engineering to aid in effective planning as well as construction of buildings. These technologies include [31]:

- ARki
- Akular AR
- AR Headsets
- AR Instructor

8.8.1 ARki

ARki is an Augmented Reality app that allows designers to create three-dimensional models and place them on top of physical surroundings or two-dimensional floor plans using a phone camera. It further allows users to upload files in addition to the ability to connect them with their respective team members to share designs and notes.

8.8.2 Akular AR

Akular AR is a cloud-based AR technology that allows users to view three-dimensional models of physical spaces using a tablet or smartphone. It can also enhance BIM models and allows on site viewing as well as for team members to send or share in real time info like IoT sensor data and BMS.

8.8.3 AR headsets

AR Headsets are the most popular AR technologies used worldwide by workers on construction sites, where AR technologies are being implemented practically. These headsets can be used with safety helmets on and allow users to view superimposed three-dimensional models and project plans without using their hands. They also enable users to gather digital measurements and incorporate them into the current models to keep them updated.

8.8.4 AR Instructor

AR Instructor is an Augmented Reality technology developed by Arvizio Company for on the job training in manufacturing and construction industries. It provides users with pre-prepared instructions based on Augmented Reality technology. Even offsite trainers are able to add three-dimensional models, documents, videos or imageries overlaid on physical surroundings of a given user to each step of the training process.

8.9 Types of AR used in civil engineering

There are several types of Augmented Reality that are commonly used in civil engineering these days [32]:
- Contour-Based AR
- Location-Based AR
- Markers-Based AR
- Projection-Based AR
- Markerless AR
- Overlay AR

8.9.1 Contour-Based AR

Contour-Based AR is an important type of Augmented Reality that uses advanced cameras to help users describe particular objects with lines to aid in specific situations. For instance, in low-visibility situations, it can be used in automotive navigation systems to help drivers drive safely.

8.9.2 Location-Based AR

Location-Based AR aims to unify three-dimensional virtual objects with the user' physical location. It uses the location and sensors of an intelligent device to place a virtual object in the envisioned location. Moreover, it allows interactive and supportive digital information related to any geographical location, which can be especially helpful for travelers or tourists in any given place by helping them understand their environment with the assistance of three-dimensional objects, etc.

8.9.3 Marker-Based AR

Marker-Based AR is a type of Augmented Reality that uses markers, also known as target pictures, to position objects in a specific given location. This type of AR requires cameras, marker tracking, image capture and processing, etc., as components to function properly.

8.9.4 Projection-Based AR

Projection-Based AR focuses on representing virtual three-dimensional objects in the user's physical space and is used to display digital data on a static background. Through this type of AR it is possible to project artificial light onto flat surfaces that exist in reality, creating illusions related to the orientation, position as well as depth of an object.

8.9.5 Markerless AR

Markerless AR is a type of Artificial Reality that analyzes properties in real-time to locate three-dimensional objects in a real image environment. It takes the help accelerometers, GPS as well as cameras in smartphones, etc., with AR software, to work properly.

8.9.6 Overlay AR

Overlay AR is a type of Augmented Reality, which has the ability to replace an object's original view from the user's point of a view with the object's updated virtual image. It allows users to see various views of the targeted object with supplementary relevant information displayed on the screen.

8.10 Methodology

In this section, the author explains the use of Augmented Reality in the field of civil engineering across different stages. Augmented Reality can be used in project management to help civil engineers manage construction projects more efficiently. It can also be helpful for engineering faculties teaching civil engineering subjects to engineering students in universities and colleges, by simplifying complex concepts. Additionally, it can be used for construction safety, to help engineers and workers on-site to understand and avoid potential life-threatening risks. In construction projects, it can aid in designing complex structures, which will also be beneficial for structural engineers. Moreover, it can also be used in traffic studies, transportation engineering, and water resource engineering projects to simplify the understanding of complicated cases as well as complex concepts as shown in Figure 8.1.

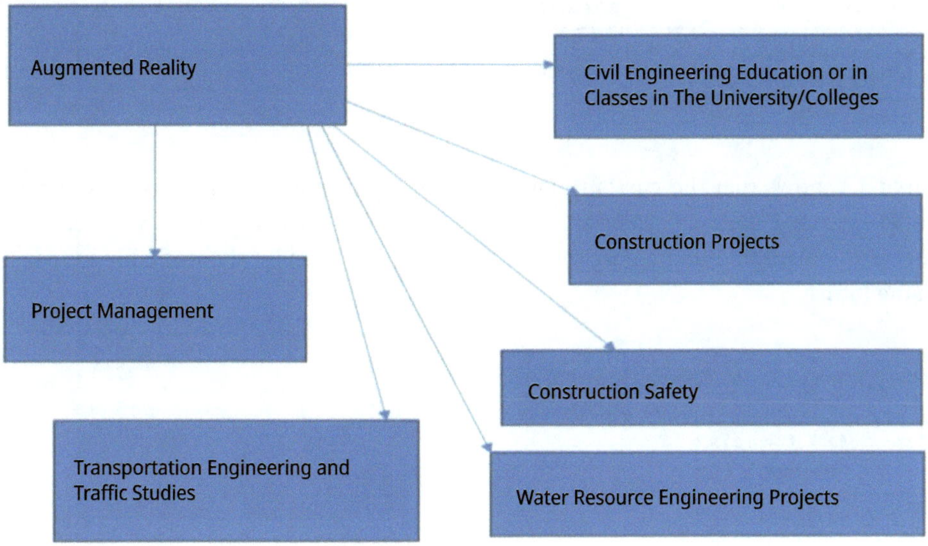

Figure 8.1: Uses of Augmented Reality in the area of civil engineering.

8.11 Graphical analysis of AR in civil engineering: Results and discussion

In this section, the author presents a graphical analysis based on data collected from various sources.

Table 8.1 shows a rapid increase of people using mobile Augmented Reality devices in the recent years and by 2024, this number is expected to increase almost by four times.

Table 8.1: Mobile AR user devices in the world [34].

Mobile AR user devices in the world	Number of users (in millions)
2019	440
2024	1730

Table 8.2 shows that only 1 % of retailers worldwide uses Augmented Reality, whereas 52 % of retailers worldwide are not ready to use Augmented Reality. This number is expected to improve in the future.

Table 8.2: Percentage of retailers using or not prepared to use AR [35].

Retailers globally	Percentage
Retailers using AR	1
Retailers not prepared to use AR	52

Table 8.3 shows that the current market size of Augmented Reality (in terms of revenue) exceeds 15 billion USD. By 2028, the market size of Augmented Reality (in terms of revenue) is expected to increase sixfold, making it a tremendously profitable industry in the future.

Table 8.3: Market size of AR (revenue) in USD [33].

Year	Market size of AR (revenue) in USD
2021	15,200,000,000
2028	90,800,000,000

Table 8.4 shows the market worth of Augmented Reality, which was of 1 billion USD in 2016, but is expected to increase by almost fifty times in eight years.

Table 8.4: AR market worth in USD [37].

Years	AR market worth USD
2016	1,000,000,000
2024	50,000,000,000

Table 8.5 shows the investment in Augmented Reality training and in industrial maintenance and compares in the year 2020, measured in USD. The investment amount in both areas is almost the same.

Table 8.5: Investment in AR in 2020 in USD [37].

Investment in 2020	In USD
AR Training	4,100,000,000
Industrial maintenance	4,100,000,000

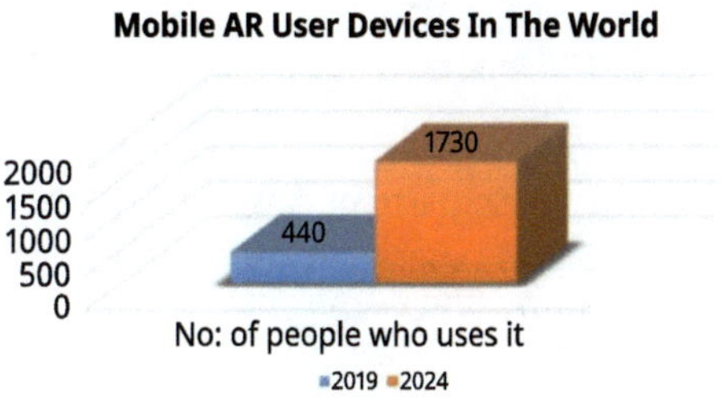

Figure 8.2: Mobile AR user devices worldwide (in millions) [34].

The above figure (Figure 8.2) illustrates the significant increase in the number of people using mobile Augmented Reality devices in recent years. By 2024, this number is expected to increase almost fourfold.

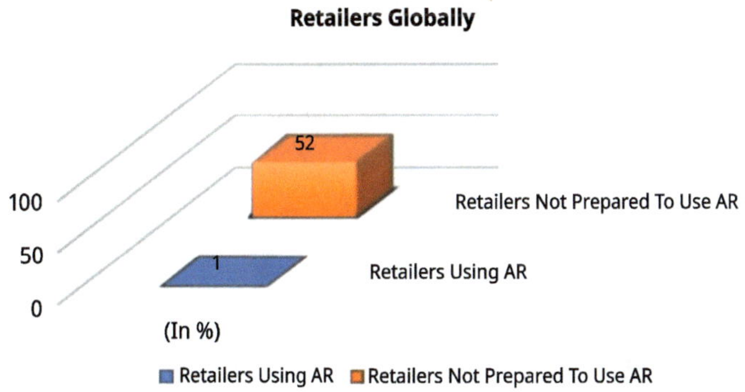

Figure 8.3: Percentage of retailers using or not prepared to use AR [35].

The above figure (Figure 8.3) shows that only one percent of retailers globally use Augmented Reality, while 52 percent are not yet prepared to use it. However, this number is expected to improve in the future.

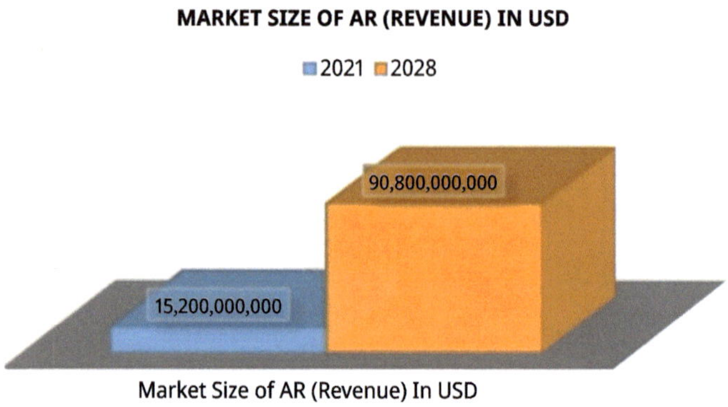

Figure 8.4: Market size of AR (revenue) in USD [33].

The above figure (Figure 8.4) shows that the current market size of Augmented Reality (in terms of revenue) is over 15 billion USD. By 2028, it is expected to increase almost sixfold,, making it a highly profitable industry in the future.

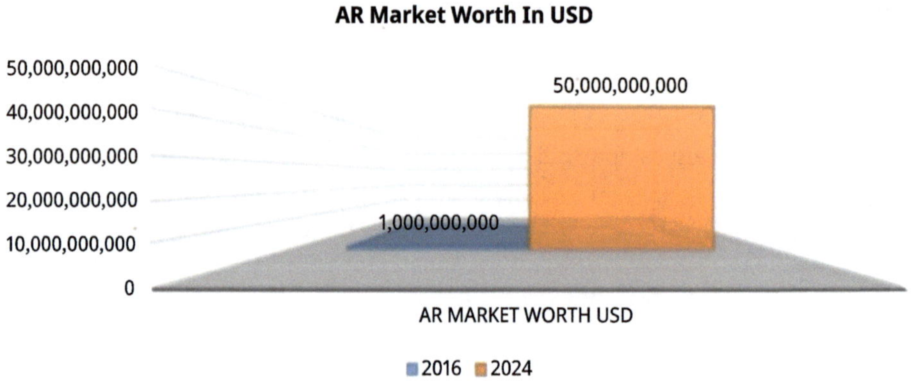

Figure 8.5: Showing AR market worth in USD [37].

The above figure (Figure 8.5) shows the market worth of Augmented Reality in USD, which was 1 billion in 2016. However, it is expected to increase almost fiftyfold within eight years.

Figure 8.6 shows the investment in Augmented Reality training and industrial maintenance in 2020, measured in USD. The investment amount is almost the same.

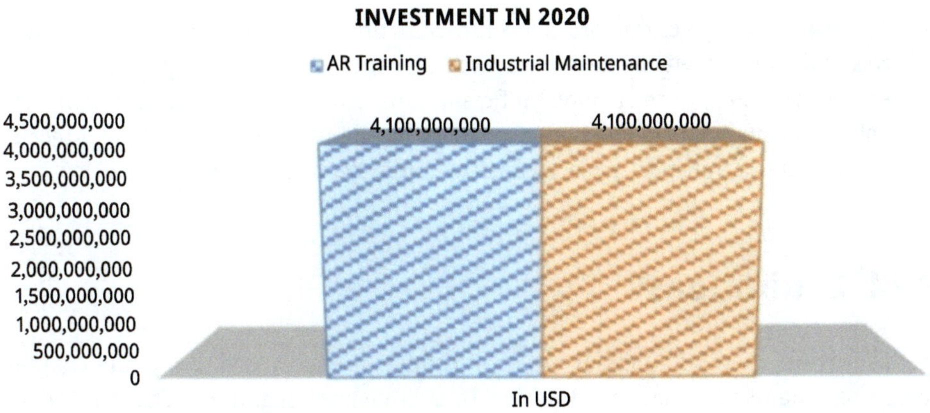

Figure 8.6: Investment in AR in 2020 in USD [37].

8.12 Advantages of Augmented Reality

Some major benefits of Augmented Reality are
- AR enables individualized learning and enhances the learning process.
- AR offers a wide range of applications that are constantly being improved.
- AR technology aids in improving efficiency as well as accuracy.
- It allows the sharing of experiences or knowledge over long distances.
- It enhances the understanding of complex concepts and makes them easier to comprehend.

8.13 Major disadvantages of Augmented Reality in the area of civil engineering

Some major drawbacks of Augmented Reality in the area of civil engineering are:
- The devices compatible with Augmented Reality are sensitive to weather changes, and bad weather can adversely affect their performance on site.
- A good internet connection is a must, and in many civil engineering projects, people barely have decent internet connection and there are places where insufficient infrastructure is available to support any internet connection.
- Civil engineers often face terrains that don't support internet connection while working on site, which can easily make Augmented Reality technology unusable on site.
- The cost of using Augmented Reality is definitely higher than the conventional ways of working on a civil engineering project.

- As a result, many contractors, subcontractors and engineers who want to reduce cost of the project will be hesitant to use Augmented Reality in the projects.
- People are also required to have sufficient skills to use Augmented Reality comfortably, and compatible devices are needed at the site while working on the projects, whether it is a transport project, construction project, and so on.

8.14 Limitations

In this section, the limitations of the book chapter are discussed. While the chapter focuses on the use of Augmented Reality in civil engineering, in the chapter, it does not explore the potential applications of Augmented Reality in other branches and sub-branches of engineering, or in general education, etc. Additionally, the author only briefly explains the concept of Augmented Reality and its uses in civil engineering, without delving into its history, evolution, or specific applications in civil engineering projects. While the major advantages and challenges of using Augmented Reality in civil engineering are discussed, they are also mentioned in brief only.

8.15 Solutions and recommendations

The book chapter covers the concept and uses of Augmented Reality, particularly in the field of civil engineering. While Augmented Reality has the potential to revolutionize the civil engineering industry, there are some major challenges that need to be addressed. There is a need to reduce the high cost of using Augmented Reality on site, a need for more advanced and compatible devices, to smoothly work with Augmented Reality in the projects as well as a need to work on increasing the awareness of the public, contractors, sub-contractors, and civil engineers to increase the acceptance of Augmented Reality. Additionally, there is a requirement of highs-speed internet access at the construction sites, in order to allow the use of Augmented Reality in the civil engineering projects without any disruption.

8.16 Future research directions

In the last decade, Augmented Reality has drawn a lot of attention, focusing on its use in different areas, such as gaming, education, medical sciences, engineering, and so on. Augmented Reality was also noticed in the field of civil engineering, but so far the practical use of AR in civil engineering has been limited. This is also due to the current technological limitations as well as the high cost of the setup, proper sophisticated facilities to support Augmented Reality and its execution on site. These issues can be addressed by future research on the use of Augmented Reality in civil engineering projects, including geotechnical engineering, transport projects, structure engineering, water resources engineering, and construction projects. Additionally, future technological advancement in Augmented Reality, a decrease in setup costs, availability of compatible devices and other required things for its smooth run, will lead to lower costs, which will encourage its use in civil engineering projects while working on site.

8.17 Conclusion

The author focused on explaining the concept of Augmented Reality as well as the use of Augmented Reality in the civil engineering field. The book chapter covered the advantages but also the challenges of using AR in civil engineering projects. Additionally, he performed a comprehensive review of numerous scientific and research papers published in the recent years about Augmented Reality in general or its use in the area of civil engineering in particular, in order to find out the gaps in the current research. The author also collected data from various sources and performed a graphical analysis, to support his study.

Bibliography

[1] F. Bellalouna, "Use Case of the Application of the Optical-See-Through Augmented Reality Technology in the Engineering Context," Procedia CIRP, vol. 106, pp. 3–8, 2022.

[2] M. Rohil and Y. Ashok, "Visualization of Urban Development 3D Layout Plans with Augmented Reality," Results in Engineering, vol. 14, 100447, June 2022.

[3] A. Settimi, J. Gamerro, and Y. Weinand, "Augmented-Reality-Assisted Timber Drilling with Smart Retrofitted Tools," Automation in Construction, vol. 139, 104272, July 2022.

[4] A. H. Behzadan, S. Dong, and V. R. Kumar, "Augmented Reality Visualization: A Review of Civil Infrastructure System Applications," Advanced Engineering Informatics, vol. 29, no. 2, pp. 252–267, April 2015.

[5] S. Alirezaei, H. Taghaddos, K. Ghorab, A. N. Tak, and S. Alirezaei, "BIM-Augmented Reality Integrated Approach to Risk Management," Automation in Construction, vol. 141, 104458, September 2022.

[6] A. Sidani, F. M. Dinis, J. Duarte, L. Sanhudo, D. Calvetti, J. S. Baptisa, J. P. Martins,and A. Soeiro, "Recent Tools and Techniques of BIM-Based Augmented Reality: A Systematic Review," Journal of Building Engineering, vol. 42, 102500, October 2021.

[7] J. M. D. Delgado, L. Oyedele, P. Demian, and T. Beach, "A Research Agenda for Augmented and Virtual Reality in Architecture, Engineering and Construction," Advanced Engineering Informatics, vol. 45, 101122, August 2020.

[8] M. Z. A. Muthalif, D. Shojaei, and K. Khoshelham, "A Review of Augmented Reality Visualization Methods for Subsurface Utilities," Advanced Engineering Informatics, vol. 51, 101498, January 2022.

[9] V. Gomez-Jauregui, C. Manchado, J. Del-Castillo-Igareda, and C. Otero, "Quantitative Evaluation of Overlaying Discrepancies in Mobile Augmented Reality Applications for AEC/FM," Advances in Engineering Software, vol. 127, pp. 124–140, January 2019.

[10] C. Kwiatek, M. Sharif, S. Li, C. Haas, and S. Walbridge, "Impact of Augmented Reality and Spatial Cognition on Assembly in Construction," Automation in Construction, vol. 108, 102935, December 2019.
[11] D. G. M. Vargas, K. K. Vijayan, and O. J. Mork, "Augmented Reality for Future Research Opportunities and Challenges in the Shipbuilding Industry: A Literature Review," In Procedia Manufacturing, pp. 497–503. #45, 2020.
[12] H. L. Chi, M. K. Kim, K. Z. Liu, J. P. P. Thedja, J. Seo, and D. E. Lee, "Rebar Inspection Integrating Augmented Reality and Laser Scanning," Automation in Construction, vol. 136, 104183, April 2022.
[13] A. Harikrishnan, A. S. Abdallah, S. K. Ayer, M. E. Asmar, and P. Tang, "Feasibility of Augmented Reality Technology for Communication in the Construction Industry," Advanced Engineering Informatics, vol. 50, 101363, October 2021.
[14] A. D'Anniballe, J. Silva, P. Marzocca, and A. Ceruti, "The Role of Augmented Reality in Air Accident Investigation and Practitioner Training," Reliability Engineering & Systems Safety, vol. 204, 107149, December 2020.
[15] M. AlFadalat and W. Al-Azhari, "An Integrating Contextual Approach Using Architectural Procedural Modeling and Augmented Reality in Residential Buildings: The Case of Amman City," Heliyon, vol. 8, no. 8, e10040, August 2022.
[16] J. Haemmerli, A. Davidovic, L. Chavaz, T. R. Meling, K. Schaller, and P. Bijlenga, "Evaluation of the Effect of Standard Neuronavigation and Augmented Reality on the Integrity of the Perifocal Structures During a Neurosurgical Approach: The Safety Task," Brain and Spine, vol. 1, Supplement 2, 100834, 2021.
[17] X. Hu, Y. M. Goh, and A. Lin, "Educational Impact of an Augmented Reality (AR) Application for Teaching Structural Systems to Non-Engineering Students," Advanced Engineering Informatics, vol. 50, 101436, October 2021.
[18] S. Gazzotti, F. Ferlay, L. Meunier, P. Viudes, K. Huc, A. Derkazarian, J-P. Friconneau, B. Peluso, and J-P. Martins, "Virtual and Augmented Reality Use Cases for Fusion Design Engineering," Fusion Engineering and Design, vol. 172, 112780, November 2021.
[19] N. Arulanand, A. R. Babu, and P. K. Rajesh, "Enriched Learning Experience Using Augmented Reality Framework in Engineering Education," Procedia Computer Science, vol. 172, pp. 937–942, 2020.
[20] Z. Wang, X. Bai, S. Zhang, M. Billinghurst, W. He, P. Wang, W. Lan, H. Min, and Y. Chen, "A Comprehensive Review of Augmented Reality-Based Instruction in Manual Assembly, Training and Repair," Robotics and Computer-Integrated Manufacturing, vol. 78, 102407, December 2022.
[21] V. Sangiorgio, S. Martiradonna, F. Fatiguso, and I. Lombillo, "Augmented Reality-Based Decision Making (AR-DM) to Support Multi-Criteria Analysis in Constructions," Automation in Construction, vol. 124, 103567, April 2021.
[22] S. Xiang, R. Wang, and C. Feng, "Mobile Projective Augmented Reality for Collaborative Robots in Construction," Automation in Construction, vol. 127, 103704, July 2021.
[23] X. Li, W. Yi, H-L. Chi, X. Wang, and A. P. C. Chan, "A Critical Review of Virtual and Augmented Reality (VR/AR) Applications in Construction Safety," Automation in Construction, vol. 86, pp. 150–162, February 2018.
[24] K. Kim, H. Kim, and H. Kim, "Image-Based Construction Hazard Avoidance System Using Augmented Reality in Wearable Device," Automation in Construction, vol. 83, pp. 390–403, November 2017.
[25] H. Pranoto and F. M. Panggabean, "Increase the Interest in Learning by Implementing Augmented Reality: Case Studies Studying Rail Transportation," Procedia Computer Science, vol. 157, pp. 506–513, 2019.
[26] F. P. Diez, J. C. Sinca, D. R. Valles, and J. M. C. Cacheda, "Evaluation of Transport Events with the Use of Big Data, Artificial Intelligence and Augmented Reality Techniques," Transportation Research Procedia, vol. 58, pp. 173–180, 2021.
[27] Z. F. Yount, S. J. Kass, and J. E. Arruda, "Route Learning with Augmented Reality Navigation Aids," Transportation Research. Part F, Traffic Psychology and Behaviour, vol. 88, pp. 132–140, July 2022.

[28] R. Lovreglio and M. Kinateder, "Augmented Reality for Pedestrian Evacuation Research: Promises and Limitations," Safety Science, vol. 128, 104750, August 2020.
[29] S. Bagassi, F. De Crescenzio, S. Piastra, C. A. Persiani, M. Ellejmi, A. R. Groskreutz, and J. Higuera, "Human-in-the-Loop Evaluation of an Augmented Reality Based Interface for the Airport Control Tower," Computers in Industry, vol. 123, 103291, December 2020.
[30] M. Smith, J. L. Gabbard, G. Burnett, C. Hare, H. Singh, and L. Skrypchuk, "Determining the Impact of Augmented Reality Graphic Spatial Location and Motion on Driver Behaviors," Applied Ergonomics, vol. 96, 103510, October 2021.
[31] BigRentzm, https://www.bigrentz.com/blog/augmented-reality-construction (last accessed 2022/12/30).
[32] Construction Placements, https://www.constructionplacements.com/augmented-reality-in-construction/#gsc.tab=0 (last accessed 2022/12/30).
[33] GlobeNewswire, https://www.globenewswire.com/en/news-release/2022/06/28/2470583/0/en/Global-Opportunities-in-Augmented-Reality-Market-Size-Will-Grow-Over-US-90-8-Billion-by-2028-at-31-5-CAGR-Growth-AR-Industry-Demand-Share-Trends-Statistics-Key-Players-Segments-Ana.html (last accessed 2022/12/30).
[34] Statista, https://www.statista.com/statistics/1098630/global-mobile-augmented-reality-ar-users/ (last accessed 2022/12/30).
[35] Threekit, https://www.threekit.com/20-augmented-reality-statistics-you-should-know-in-2020/ (last accessed 2022/12/30).
[36] Wikipedia, https://en.m.wikipedia.org/wiki/Augmented_reality/ (last accessed 2022/12/30).
[37] Data Prot, https://dataprot.net/statistics/augmented-reality-stats/ (last accessed 2022/12/30).
[38] TeamViewer, https://www.teamviewer.com/en-us/augmented-reality-ar-vs-virtual-reality-vr/ (last accessed 2022/12/30).
[39] TeamViewer, https://www.teamviewer.com/en-us/augmented-reality-in-construction/ (last accessed 2022/12/30).

Ajay Sudhir Bale, Muhammed Furqaan Hashim, Kajal S Bundele, and
Jatin Vaishnav

9 Applications and future trends of Artificial Intelligence and blockchain-powered Augmented Reality

Abstract: Artificial Intelligence (AI) is a sophisticated computer system designed to simulate human intelligence under different circumstances. It incorporates several branches of study and utilizes existing technologies to demonstrate human behavior and autonomously take necessary actions according to any particular situation. A more recent innovation in capturing and storing data is the blockchain technology, which shares data across several nodes in a distributed network. Augmented Reality (AR) is a technology that aims to provide users with an interactive experience of their real-world environment by enhancing real-world objects through computer graphics and superimposing virtual computer-generated images on a user's view of the real world. Implementation of AI further enhances the experience of AR by integrating its abilities and adding new features to the AR intelligence, such as text analysis, object detection, scene labeling, and much more. It enables AR to interact with the physical world using neural networks. This paper focuses on exploring the different applications of AI for AR and how fusing blockchain with them can support the entire AR ecosystem.

9.1 Introduction

For centuries, humans have utilized their knowledge of science and mathematics to invent and explore new and improvised ways of working more efficiently. This constant thirst for exploration has led to significant innovation in existing systems and sparked various technological and industrial revolutions [1–3]. These revolutions significantly advanced scientific research and brought about numerous changes in day-to-day processes. More recently, the focus of innovation has shifted to computation and virtualization, owing to its great value and its benefits. As the world moves into the sphere of virtualization, Augmented Reality has found its way into the spotlight, with more and more applications using it in several different ways to change the user experience. With

Ajay Sudhir Bale, Department of ECE, New Horizon College of Engineering, Bengaluru, India, e-mail: ajaysudhirbale@gmail.com
Muhammed Furqaan Hashim, Kajal S Bundele, Jatin Vaishnav, Department of CSE, School of Engineering and Technology, CMR University, Bengaluru, India, e-mails: furqaanhashim@gmail.com, kajalbundele26@gmail.com

https://doi.org/10.1515/9783110785234-009

the incorporation of artificial intelligence technology, the capabilities of Augmented Reality can be enhanced, making it stronger and broadening its use cases.

Augmented Reality (AR) is a technology that aims to provide users with an interactive experience of their real-world environment. [4] It is achieved by enhancing objects in the real world through computer graphics and superimposing virtual computer-generated images on a user's view of the real world as seen in Figure 9.1. It creates an illusion, making the user believe that the virtual objects are present in their environment within the real world. [5] Users can experience AR and bring their environment to life using AR devices such as optical see-through or video see-through head-mounted displays, virtual retinal systems, and monitor or projector-based displays, as seen in [6].

Figure 9.1: Concept of Augmented Reality.

Figure 9.1 depicts how AR superimposes computer-generated images on the physical environment of users. In this case, a dinosaur has been virtually superimposed with the help of a smart device. The roots of Artificial Intelligence, often known as AI, can be traced back to the mid-20th century. The fundamental base for robotics, Asimov's Laws, first appeared in a short science fiction story, Runaround, written by American science fiction writer Isaac Asimov in 1942. [7] AI is a sophisticated computation system designed to simulate human intelligence under different circumstances. It incorporates numerous branches of study, as shown in Figure 9.2, and utilizes existing technologies to demonstrate human-like behavior and take necessary actions autonomously according to any particular situation faced by the system.

Blockchain technology was introduced in the early 1990s [10] and sparked great interest in computer science enthusiasts, who worked on fine-tuning and expanding its functionality over years of extensive research and development. It gained fame after an anonymous person or group of people created Bitcoin [16], the first-ever digital cryptocurrency [15]. Blockchain is a revolution in data management and storage, offering decentralization capabilities since the data is stored across a distributed database. In the blockchain system, data blocks containing information are generated and appended to a chain of blocks; [11–14]. the nodes on the blockchain network digitally sign and verify every transaction taking place, thereby eliminating verification from a centralized system and ensuring decentralization across the chain [15, 17, 18].

The upcoming chapters of this paper attempt to understand how AI and blockchain fit into the picture of AR technology and explore the possible ways in which they can enhance its experience, making it intelligent, broadening its use cases, and expanding its utility. This study also explores the different industrial areas where the fusion of the three technologies can improve work lifestyle and user experiences. AI derives concepts from different areas of science to create a computation system that can demonstrate human behavior. As shown in the Figure 9.2, these branches of study formulate the foundations of AI.

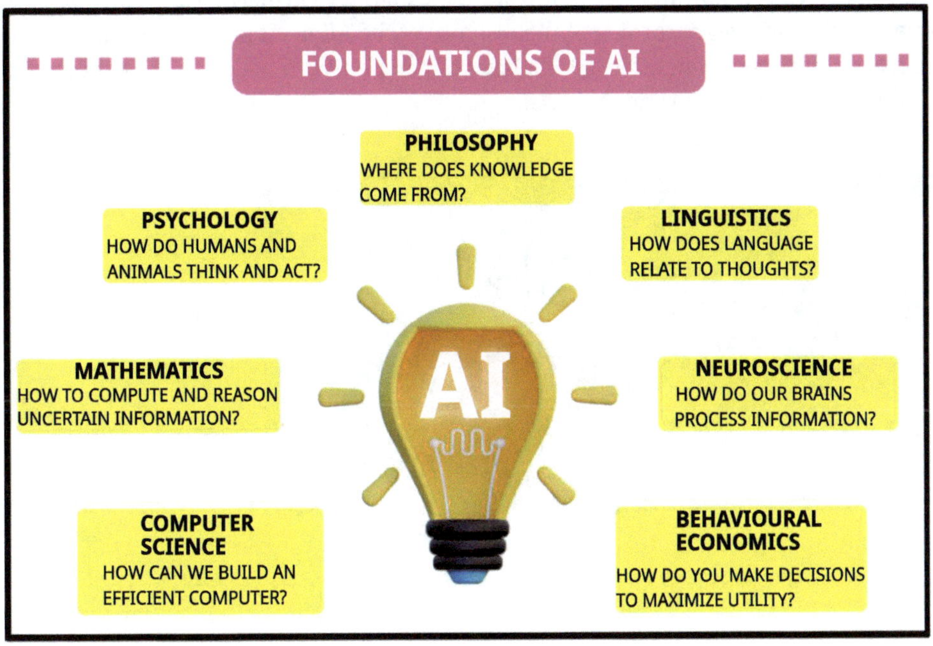

Figure 9.2: Foundations of AI.

9.2 Methodology

The assessment of this study involved a scientometric analysis of bibliographic data available on the three technologies in focus—AR, AI, and blockchain. To gain a comprehensive understanding of the three technologies, we conducted an extensive review of the literature, including research papers, articles, scientific journals, and credible websites. In the absence of available texts on the applications and future trends of AI and blockchain-powered AR technology, we drew connections between the three technologies based on our analysis of their capabilities and potential applications.

9.3 How AI and blockchain transformed AR

AR technology enables users to have an interactive and personal experience with real-world objects by bringing them to life with the help of computer graphics. Graphical images virtually superimpose over objects in a user's physical environment, which they may view using an AR display device. When a user points the AR device to any physical object in the real world, the AR software perceives and studies it using computer vision [5, 8] to derive relevant information from the cloud. [9] Furthermore, virtual objects tune and change sizes according to the distance of the AR device and the environment where it is projected. This effect adds a layer of perspective to the entire AR experience as shown in Figure 9.3 (a) and (b).

(a) (b)

Figure 9.3: Change in image size based on perspective.

Figure 9.3 (a) and (b) illustrate how virtual objects adjust in size according to the distance between the user and the projected object. In (a), the object is superimposed closer to the user, making it appear more prominent in size. When the object moves farther away in (b), it shrinks in size, similar to real-world objects. By fusing AI and blockchain technology with AR, users experience the best of the three worlds. The various fields of AI depicted in Figure 9.4 help create a more intuitive and intelligent AR system that can do more than just draw virtual objects onto the physical environment. It can scan the environment to analyze and process it with its cognitive computing ability, providing users with a more intuitive experience. Furthermore, AI-powered AR systems can learn

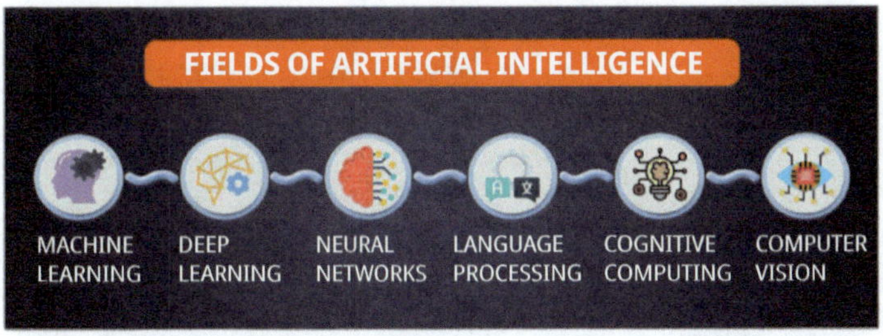

Figure 9.4: Fields of Artificial Intelligence.

users' choices and preferences using machine learning and deep learning to provide personalized solutions.

Different technologies converge into the formation of AI, making it a sophisticated innovation and the culmination of years of research and development, as shown in Figure 9.4. On the other hand, blockchain offers unparalleled services, including enhanced security, greater transparency, improved traceability, increased efficiency and speed, and an intelligent automation system [19]. AR technology can significantly benefit from these unique features. It unlocks numerous opportunities for implementing features such as utilizing digital assets or NFTs [20, 21], secure transactions, data transparency, and much more in the AR world. These features significantly improve AR technology and enable it for a wide range of applications, as seen in Section 9.4.

9.4 Applications of AI and blockchain-powered AR

9.4.1 Healthcare

The findings in [22] show that large clinics are currently utilizing AI-enabled technologies to assist medics in patient assessment and treatment actions for a variety of disorders. AI technologies are also having an influence on how well clinics handle their supervisory and medical staff. It is evident that the rapid development of AI and technological applications would assist healthcare providers in enhancing patient value and streamlining control activities. One of the key response methods for COVID-19 and other advancing viral infections is the quick creation and use of low-cost blockchain and AI-coupled techniques, as depicted in Figure 9.5 [23].

This approach involves using a mobile device or tablet program as the first step. The program can be modified from existing self-testing apps and will prompt the user for their unique identity before requesting their medical reports. Thanks to the blockchain and AI system, epidemic tracking officials will be informed of all testing, including the number of positive and negative results. This will allow for positive patients to be sent to an isolation facility for care and observation. Built-in geographic information systems

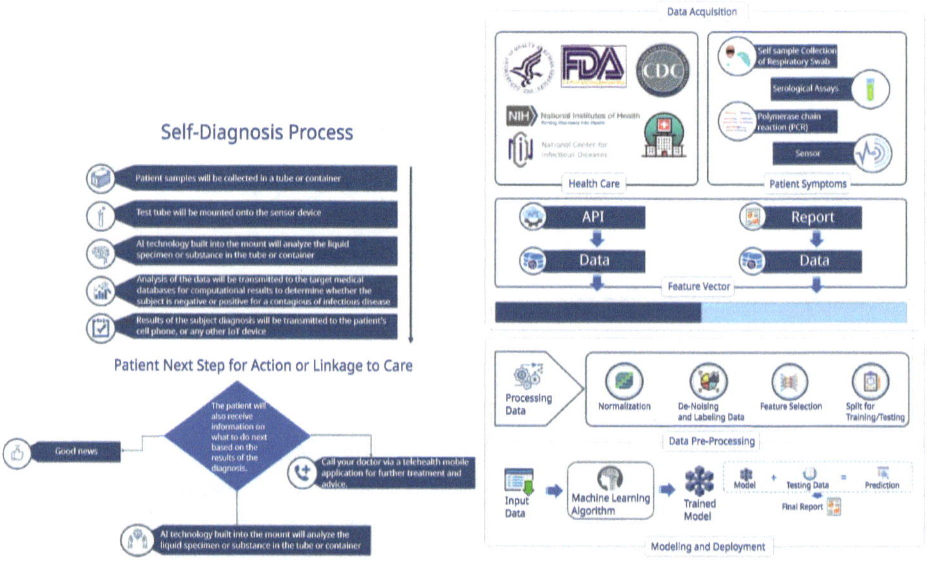

Figure 9.5: Mobile-linked self-testing and monitoring solution for detecting serious diseases using blockchain and AI [23]. https://www.mdpi.com/2075-4418/10/4/198.

(GIS) on cellular phones will make it possible to track those who tested positive. To guarantee adequate monitoring and epidemic management, this platform will be linked to regional and global networks.

The AI component of the system will enable powerful information gathering (patient data, location, and lab tests), analysis, and collection of data from confederated blockchain systems to derive triangulated data at extremely high levels of trust and speed. This guarantees secure and unchangeable data sets with a carefully designed unified software system, which allows the gathering of high-quality information and drawing of insightful conclusions. The growth of such tests can help overcome supply chain difficulties [24] and price barriers that may prevent point-of-care (POC) tests from being accessible in areas with limited resources. With the challenges faced by covid-19 epidemic, this method could be adapted for community-based incidence discovery for illnesses like HIV, TB, and malaria.

Metaverse is a web platform that facilitates communication between users and their surroundings. Health organizations can use Metaverse software to provide dental education and interactions to specific groups and communities, regardless of location, implying no travel expenses [25]. In [25], it is also reported that harmful practices and commodities are promoted through internet ads, which is a significant global industry. Authorities and politicians are thus asked to implement stringent regulations to prevent these ads for harmful products on the Metaverse platform, particularly those for hard liquor and tobacco, from negatively impacting the user's health. The Metaverse network is anticipated to promote dental health, a healthier diet, and comfort as guiding principles. The first AR procedures on living patients were carried out by Johns Hopkins neurosurgeons [26]. An eye display headgear was used, which projected photos of the person's anatomy, including joints as well as other tissues, based on CT scans, giving the doctors an x-ray vision as reported in [27].

Figure 9.6 illustrates the potential for AR technology to assist in diagnosing internal body parts. In the coming years, dentistry will also advance by learning from patient medical practices as they explore the Metaverse. It may soon be possible to conduct dentistry telemedicine chats with clients through virtual characters in a virtual metaverse. One can envision performing dental procedures while viewing real x-ray or 3D images of the canal's geometry, inserting an implant while observing the its precise location relative to the bones during surgery, or eliminating a cancer growth while viewing a camera stream of the tumor's physical expansion. The blockchain technology ensures the immutability of recorded transactions or network agreements in a backward-incompatible manner. Every recorded transaction is regarded as money (such as Bitcoin), and contains a secure electronic signature. Smart contracts, including "tokens" and "NFTs" (non-fungible tokens), were created based on this technology. Immediate transactions are paid, and small trading costs are applied. As a result, medical professionals may use cryptocurrencies to pay for physical and digital services and products [28].

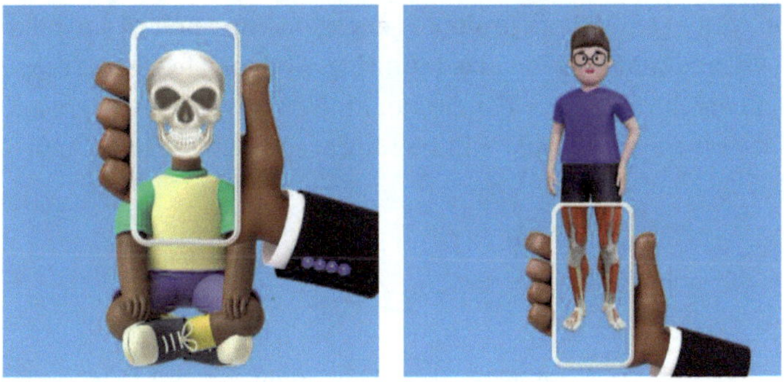

(a) (b)

Figure 9.6: Illustration of diagnosing internal body parts using AR devices.

9.4.2 Education

Many academics have investigated the possible applications of AI and blockchain in academia, with a focus on the benefits of these innovations in enhancing children's educational opportunities and outcomes. These innovations can be used to ensure equal educational opportunities for almost all learners, including those with impairments, refugees, and those living in remote or rural locations. AI-powered virtual objects and bots can help children with special needs return to school from home or from a clinic, ensuring educational continuity during times of crisis, such as the COVID-19 pandemic

[29]. Additionally, educators and professors can use these cutting-edge tools to conduct adaptive chat sessions, increase participant engagement with intelligent tutoring systems, make remote group conversations more engaging, in conjunction with e-learning, and facilitate simple assessment and assignment marking in larger groups, while also providing measures to prevent cheating. The use of these tools in education is still in its early stages, and much needs to be examined, evaluated, and communicated. Nevertheless, many universities and organizations worldwide are experimenting with new AI and blockchain applications to enhance students' educational objectives.

Blockcerts is a collection of fully accessible programs, frameworks, and modules for creating and validating identities on the Bitcoin blockchain. It is considered as the first notable example of holding cryptographic credentials [30]. Blockcerts provides a quick and easy way to create a certificate with the necessary details, and they are verified with the lender's encryption key. A secret key and cert are combined to create a hash, which is then recorded in the blockchain and used to identify the recipient. For the sake of productivity, certificates can be produced in batches. The recipient (in this case, the learner) can obtain and distribute their certificates in real-time through a smartphone app called the Blockcerts Wallet [30–32].

The training technique used in [33] helped children enhance overall fluency and abilities while also expanding their language. Robot-assisted language learning (RALL), when used in conjunction with AI and robots, promoted engagement, captured and held students' interest, and delivered more personalized tutoring. Although learners recognized the importance and benefits of using VR and robot innovations in vocabulary acquisition, they believed that the system could still be improved to create many more living things.

9.4.3 Gaming

AlphaGo, AlphaGo Zero, and AlphaZero, which successively appeared and excelled in one of the most difficult games, Go, show the potential of deep learning. Deepmind's AlphaGo Zero demonstrated in 2017 that machines are capable of outperforming humans in Go without any prior knowledge or experience [34]. AlphaZero afterwards achieved great success in the games of shogi and chess, as outlined in Figure 9.7. These achievements required a significant amount of computing capabilities. Facebook's ELF OpenGo utilized 2000 V100 GPUs to reach the highest level of performance, whereas Deepmind needed 5000 TPUs to run the Go training process for several days [35]. Numerous academics, who are engaged in game AI but do not wish to spend a lot of time studying the intricate game mechanics, are finding the game of NoGo to be a beneficial tool. They may easily create the NoGo game AI's fundamental structure using the NoGoZero+ solution and obtain more time for research. As a result, the gaming AI field gains from the reduced study criteria thanks to NoGoZero+'s methodologies as shown in Figure 9.8.

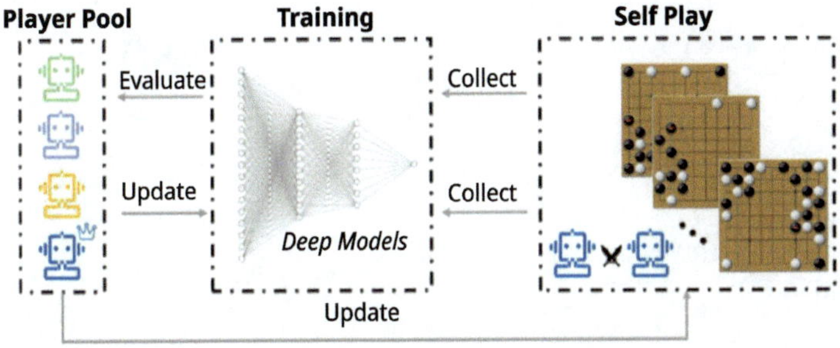

Figure 9.7: AlphaZero's pipeline [34]. https://www.mdpi.com/2079-9292/10/13/1533.

Additionally, AlphaZero-based techniques mostly employ convolutional neural networks (CNN) to gather board-related data, similar to how CNN gathers data from images in computer vision tasks. This implies that computer vision techniques can be repurposed for AI education. The basic CNN architecture can be merged with the attention mechanism, which is frequently used to pick out the more important details from images throughout computer-vision activities, to improve the efficiency of AI systems while minimizing hardware consumption.

The growth of AI in the video game business has been greatly aided by high-performance processors, but it is essential to develop the optimal assessment method. A Chinese Ludo program is shown in Figure 9.8 [36]. The optimization algorithm of the software is merely a linear combination of the values of the four model parameters, unlike the majority of chess algorithms that rely on great computer efficiency. The risk

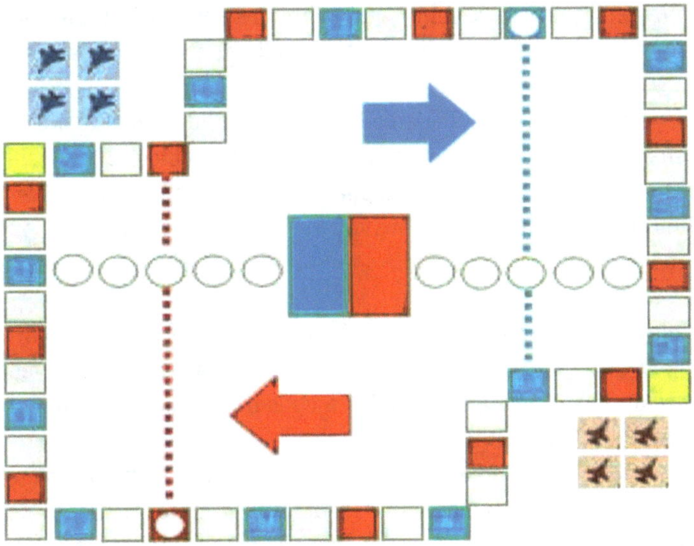

Figure 9.8: Chinese Ludo Game [36]. https://www.mdpi.com/2079-9292/11/11/1699.

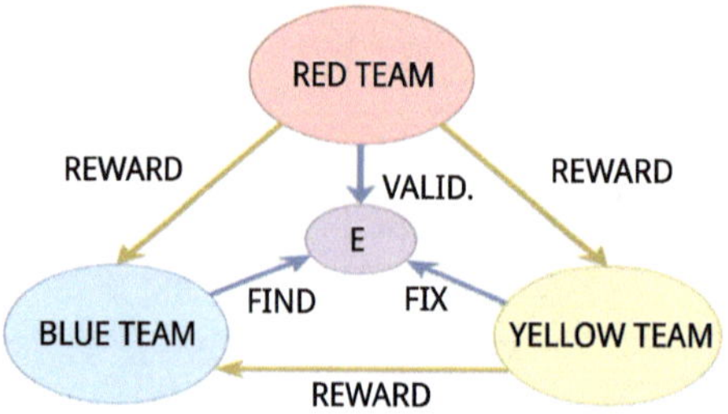

Figure 9.9: Structure of Serious Game [37]. https://www.mdpi.com/2076-3417/11/4/1449.

between the pair dice on any two positions can be quickly determined using a threat matrix, which was creatively developed as a part of this research.

Serious games as shown in Figure 9.9 [37] are those designed with a primary objective other than amusement, such as data retrieval or active engagement [38]. They are gaining popularity as a hopeful instructional tool in various fields, including the military, education, world affairs, management, and technology. In the field of medicine, serious games are receiving increasing focus due to their potential to boost user interaction, adjust to the patient's level of expertise, and provide reproducibility and ongoing learning. The use of virtual reality (VR) methods in the development of serious games to produce immersive experiences has gained popularity in the last few years. The most popular methods of using VR/AR in the healthcare industry include surgery scheduling and simulation-based clinical teaching, treating pain, patient and expert training, mental and motor rehab, surgery modeling and other medical fields. The use of mixed reality (MR), which combines real and virtual reality (VR) material to create new settings where actual and virtual items can communicate with one another in real-time, has recently increased significantly in the health industry. Applications for both VR and AR are included in MR.

The instructional game presented in [39] is an educational tool designed to aid in understanding the consensus protocol, a fundamental concept behind blockchain technology. Groups of two parties are formed in the initial phase, or each participant receives a deck of cards. Six white cards are placed face down by the teams. Each individual on the team calls out any of the crypto symbols while simultaneously turning over each card of the other team.

The corresponding player may grab a card from the other team if the cryptocurrency symbol on the flipped card matches the sign called out. Blue cards are employed to continue the game once all the white cards have been consumed. The instructional tool used in the game is shown in Figure 9.10 [39].

The fundamental rules of the game remain the same, except that at each turn, players must call out a crypto symbol and two cards from the opposing team should be flipped over. The opposing player's two cards can be grabbed when both signs align. Last but not least, three red cards from the other team can only be claimed if three

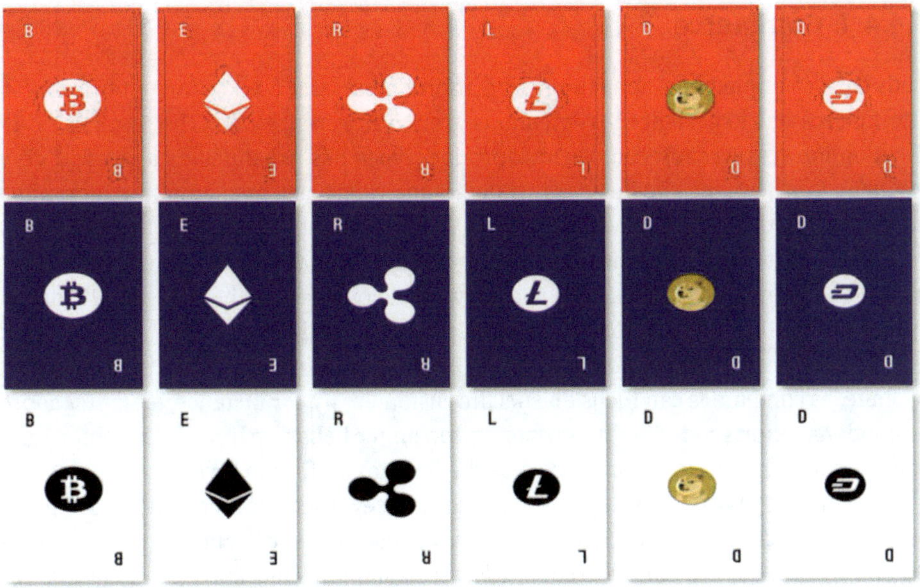

Figure 9.10: Instructional symbols employed in this game [39]. https://www.mdpi.com/2076-3417/11/4/1449.

cryptocurrency symbols match. The winner is the player who wins two out of the three rounds. Learners can improve their understanding of cryptocurrency concepts by engaging in educational games like this one. The gaming approach used to convey fairly complex ideas to primary school students through educational games was confirmed to be effective after educational experts provided input on the designed educational intervention.

The CryptoKitties game highlighted some of the of the limitations of blockchain technology [40]. Due to the popularity of CryptoKitties (more than 40,000 daily active users), the Ethereum network received extra traffic, which caused the number of awaiting entries on the Ethereum blockchain to multiply by six [41]. The functionality of these games is also plagued by a number of additional problems for the creators, including extended lags, development problems, and last but not last, expenses. Consequently, creating blockchain-based games is not an easy undertaking and can occasionally be quite challenging, resulting in unintuitive solutions. In [39], the study examined how unique portable gaming environments and improved in-game functionality, such as in-app purchases outside private entities and tailored, context-aware marketing, could be enabled and expanded by Distributed Ledger Technologies (DLTs) and IoT devices (mainly beacons). Additionally, the study also detailed a particular concept of a location-based mobile game that assessed it in terms of several different criteria using predetermined Key Performance Indicators (KPIs).

9.4.4 E-Commerce

With ease of implementation, simplicity, affordability, and a wide range of products, e-commerce has experienced tremendous growth in recent years. The number of diverse online retailers offering various product types has expanded as a result. Every company must enhance its business strategies in order to succeed amid the increasing number of participants. Customer Lifetime Value (CLV) is a standard metric that cross-online merchants typically consider for competition as it allows companies to identify their best customers [42]. For multi-category online businesses, the study in [42] developed a new 360° paradigm for CLV prediction, enabling them to handle all aspects of CLV forecasts and customer categorization. Providing a solution for this issue can boost revenue, as companies can focus on specific customers by, for instance, assigning more promotional efforts to individuals. Data on consumer behavior from a particular company was used and four factors served as the foundation for the investigation.

The first step was to develop the suggested framework to foresee other possibilities beyond CLV, such as a Distinct Product Category ratio (DPC) and Trend in Average Spending (TAS). Although CLV can be used to forecast a company's most profitable clients, this metric is inadequate for companies that sell products across multiple categories. As a result, we recommend forecasting DPC and TAS in addition to CLV to improve CLV predictive performance and more accurately segment clients of multi-category e-commerce businesses.

The second stage involved selecting the most effective method for estimating CLV for a multi-category company. To obtain estimates for multiple factors, a multi-output deep neural network (DNN) model was developed to use the same parameters to anticipate various outcomes.

The third concerns interpreting the outcomes of the multi-output DNN model to clarify the specifics of the predictive value and increase consumer trust in DNN models. The final step involved grouping customers based on the output attributes that were suggested: DPC, TAS, and CLV (rather than using only CLV).

Based on the service meet hypothesis and the superposition theory, the study in [43] conducted two purchasing experiments to record consumers' emotions in order to assess the impacts of three different online support models on consumers' purchasing behavior (AI customer service, manual customer care, and human-machine cooperation service quality). The findings show a positive relationship between customer satisfaction and the likelihood that they will buy a product. Additionally, the findings demonstrate a mediator role for the type of item in the connection between online customer interaction and purchase intention. The model is illustrated in Figure 9.11.

The long-term viability of using e-commerce to increase life satisfaction depends on various factors. The acceptance of electronic commerce is currently influenced by several crucial elements, including trust and payment options. Consequently, the study in [44] focuses on the functions of trustworthiness and form of payment to examine the Unified Theory of Acceptance and Use of Technology (UTAUT) drivers of consumer

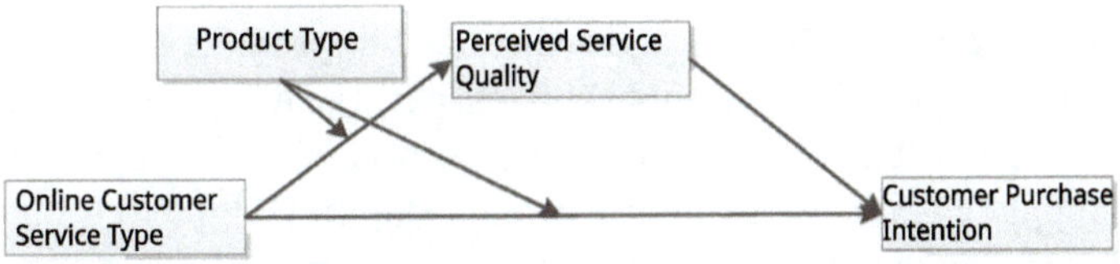

Figure 9.11: Framework used in [43]. https://www.mdpi.com/2071-1050/14/7/3974.

e-commerce adoption in Ghana. As part of the study, 535 purposeful questionnaire techniques were employed in six Ghanaian regions, and the completed questionnaires were analyzed using the Partial Least Square Structural Equation Model (PLS-SEM). The findings support the notion that e-commerce adoption and the UTAUT factors are closely related. The payment platform, however, does not appear to have any moderating impact on the relationship between e-commerce adoption and trustworthiness. Trust, on the other hand, highly mediates the adoption of online shopping and the UTAUT factors. The crossover between social influence and trust was strongest, while the crossover between performance expectancy and trust was weakest. The framework used in [44] is shown in Figure 9.12.

Decentralized forms of payment have become increasingly widespread, leading to a surge in new digital currencies entering the market. For each cryptocurrency, specific sites have been established, where users can access data and equipment for mining digital currencies on a daily basis. Users access these cryptocurrency sites through either

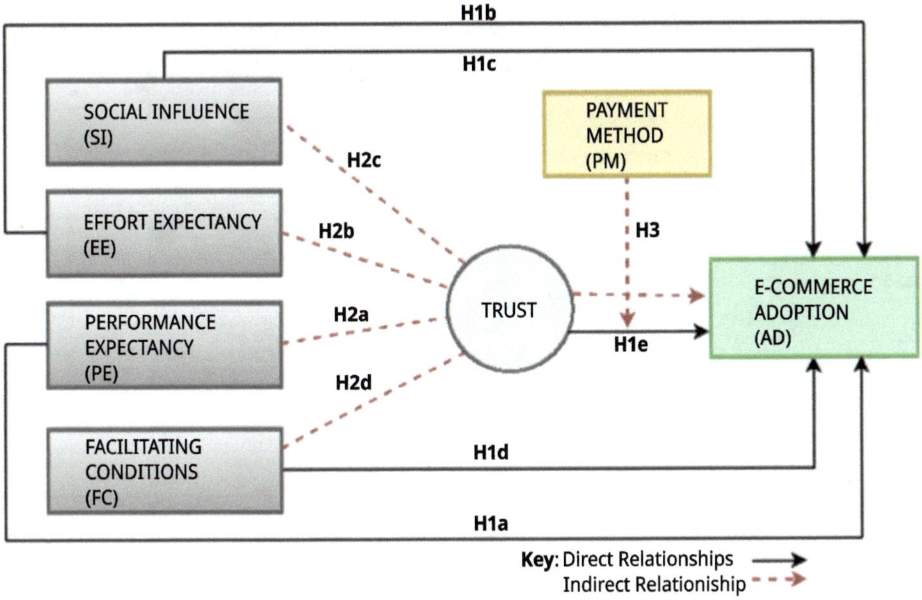

Figure 9.12: Framework used in [44]. https://www.mdpi.com/2071-1050/14/14/8466.

mobile or desktop devices. As a result, there is a growing need to promote the quality of crypto services and the consumer aspects that influence it. The aforementioned procedure increases the online visibility of crypto businesses, necessitating enhanced customer relations and supply chain strategy improvements [45]. In [45], 180 days of on-site online marketing research data were collected from 10 well-known crypto sites for both mobile and desktop platforms. As a result, a three-stage model was used. The model's first stage involves statistical and regression analysis of cryptocurrency analytics tools, which is preceded by the implementation of fuzzy cognitive mapping and agent-based models. This study examines specific web metrics and device preferences to determine strategies for advertising crypto sites. According to research findings, web analytics present prospects for further website optimization through increased internet traffic and improved online presence by providing a more comprehensive understanding of customer behavior in crypto sites.

9.4.5 Metaverse

Metaverses result from years of research and development across several different technologies, incorporating numerous disruptive innovations such as AI, AR, VR, and blockchain. They create a virtual environment to connect people's imagination to the real world and use a variety of existing technologies to build a blended environment, allowing people to interact with others through virtualized characters. [46]

In recent times, metaverses have gained significant interest among tech enthusiasts as the latest revolution in the computing industry. They generated a considerable buzz and gained widespread attention in late 2021 and early 2022. This rise in popularity can be seen in Figure 9.13, obtained from Google Trends data [47]. This sudden

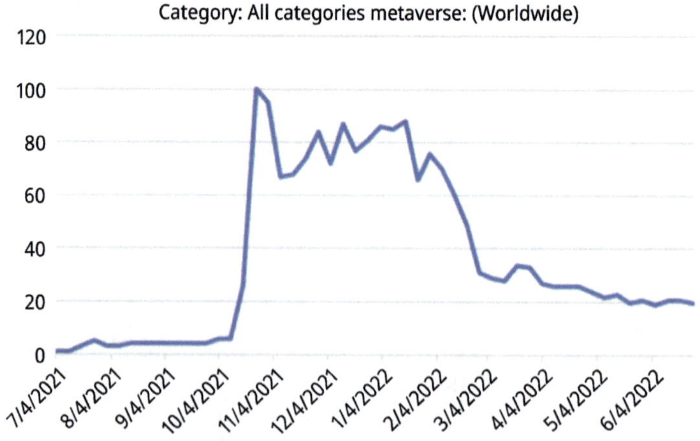

Figure 9.13: The volume of "Metaverse" queries made on Google over the previous twelve months. This information may be found in [47] and was acquired from Google Trends Data.

increase in popularity can be attributed to many companies, including Facebook and Microsoft, announcing their intentions to develop metaverses. Moreover, the concept of metaverses promises nu benefits for its users. [46] also presents several possibilities of how metaverses can transform existing systems in different industries, providing users interactive and personalized experiences.

Apart from gaming, one of the many applications of the metaverse is its ability to provide a comprehensive and collaborative virtual work experience. Since the COVID pandemic forced companies to explore the virtual realm for collaborative working, the concept of holding meetings and conferences and working virtually from home has risen in popularity. As a result, many companies began exploring the idea of developing metaverses for work. With the help of AR technology, ideas can be shared across the board more efficiently and clearly, as users can unfold their imagination and allow others to visualize their ideas using AI and graphically generated 3-D objects. The sharing of ideas and discussions can be transformed with metaverse technology. Moreover, with the inclusion of blockchain technology, the security of data and privacy and be strengthened to provide more reliability and transparency for organizations dealing with sensitive information.

9.5 Concluding Remarks

In this literature review, we have studied and presented an overview of how AI and blockchain can power AR to further broaden its use cases and provide advancements and technological solutions to numerous industries. AR enables people to experience and see virtually generated objects superimposed onto the real world. This essentially means that people can bring their imagination to life, leaving the door open for endless possibilities. Many industries have been quick to adopt this technology in order to improve their systems of work. While AR has already established its presence, research on improving and enhancing the technology is underway to better suit users' needs. This includes incorporating other established technologies such as AI and blockchain to increase AR technology's power and broaden its uses. We have seen several industries, such as healthcare, education, gaming, e-commerce, and metaverse, where the fusion of these technologies has shown great potential or has already been implemented to provide a better experience and improve work efficiency. As we continue to invest more and more in virtualization and automation to make our lives easier, it will be interesting to see where technology stands a few years down the line.

Bibliography

[1] P. Hudson, "The Industrial Revolution," Bloomsbury Publishing, 2014.
[2] A. P. Usher, "A History of Mechanical Inventions," Revised edition, Courier Corporation, 2013.

[3] M. Gibbons and R. Johnston, "The Roles of Science in Technological Innovation," Research Policy, vol. 3, no. 3, pp. 220–242, 1974. ISSN 0048-7333, https://doi.org/10.1016/0048-7333(74)90008-0.
[4] Y. Chen et al., "An Overview of Augmented Reality Technology," Journal of Physics. Conference Series vol. 1237, no. 2, IOP Publishing, 2019.
[5] A. B. Craig, "Chapter 2 – Augmented Reality Concepts," In A. B. Craig, editor, Understanding Augmented Reality, pp. 39–67, Morgan Kaufmann, 2013, ISBN 9780240824086, https://doi.org/10.1016/B978-0-240-82408-6.00002-3.
[6] R. Silva, J. C. Oliveira, and G. A. Giraldi, "Introduction to Augmented Reality," National laboratory for scientific computation 11, pp. 1–11, 2003.
[7] M. Haenlein and A. Kaplan, "A Brief History of Artificial Intelligence: On the Past, Present, and Future of Artificial Intelligence," California Management Review, vol. 61, no. 4, pp. 5–14, Aug. 2019, https://doi.org/10.1177/0008125619864925.
[8] N. Zheng, G. Loizou, X. Jiang, X. Lan, and X. Li, "Computer Vision and Pattern Recognition," International Journal of Computer Mathematics, vol. 84, no. 9, pp. 1265–1266, 2007, https://doi.org/10.1080/00207160701303912.
[9] W. A. Hoff, K. Nguyen, and T. Lyon, "Computer-Vision-Based Registration Techniques for Augmented Reality," In Proc. SPIE 2904, Intelligent Robots and Computer Vision XV: Algorithms, Techniques, Active Vision, and Materials Handling, (29 October 1996), 1996, https://doi.org/10.1117/12.256311.
[10] Icaew.Com. "History of Blockchain." https://www.icaew.com/technical/technology/blockchain-and-cryptoassets/blockchain-articles/what-is-blockchain/history (accessed 1 June 2022).
[11] Y. Liu and P. Tang, "The Prospect for the Application of the Surgical Navigation System Based on Artificial Intelligence and Augmented Reality," In 2018 IEEE International Conference on Artificial Intelligence and Virtual Reality (AIVR), pp. 244–246, 2018, https://doi.org/10.1109/AIVR.2018.00056.
[12] L. Tatwany and H. C. Ouertani, "A Review on Using Augmented Reality in Text Translation," In 2017 6th International Conference on Information and Communication Technology and Accessibility (ICTA), pp. 1–6, 2017, https://doi.org/10.1109/ICTA.2017.8336044.
[13] S. Benkerzaz et al., "A Study on Automatic Speech Recognition," Journal of Information Technology Review, vol. 10, no. 3, pp. 77–85, 2019, https://www.academia.edu/43773934/A_Study_on_Automatic_Speech_Recognition.
[14] H. Ibrahim and A. Varol, "A Study on Automatic Speech Recognition Systems," In 2020 8th International Symposium on Digital Forensics and Security (ISDFS), pp. 1–5, 2020, https://doi.org/10.1109/ISDFS49300.2020.9116286.
[15] J. Ghosh, "The Blockchain: Opportunities for Research in Information Systems and Information Technology," Journal of Global Information Technology Management, vol. 22, no. 4, pp. 235–242, 2019, https://doi.org/10.1080/1097198X.2019.1679954.
[16] B. Segendorf, "What is Bitcoin?" Sveriges Riksbank Economic Review 2014, pp. 71–87, 2014. http://archive.riksbank.se/Documents/Rapporter/POV/2014/2014_2/rap_pov_1400918_eng.pdf.
[17] A. Hayes, "Blockchain Explained," Investopedia, 13 June 2014, https://www.investopedia.com/terms/b/blockchain.asp.
[18] M. Pilkington, "Blockchain Technology: Principles and Applications," In Research Handbook on Digital Transformations, Edward Elgar Publishing, 2016, https://doi.org/10.4337/9781784717766.00019.
[19] "Benefits of Blockchain," Ibm, https://www.ibm.com/in-en/topics/benefits-of-blockchain (accessed, 9 June 2022).
[20] Y. Nagpal, "Non-Fungible Tokens (NFT's): The Future of Digital Collectibles," International Journal of Law Management and Humanities, vol. 4, no. 5, pp. 758–767, 2021, https://doij.org/10.10000/IJLMH.111984.
[21] H. Bao and D. Roubaud, "Non-Fungible Token: A Systematic Review and Research Agenda," Journal of Risk and Financial Management, vol. 15, no. 5, pp. 215–224, 2022, https://doi.org/10.3390/jrfm15050215.

[22] D. Lee and S. N. Yoon, "Application of Artificial Intelligence-Based Technologies in the Healthcare Industry: Opportunities and Challenges," International Journal of Environmental Research and Public Health, vol. 18, p. 271, 2021, https://doi.org/10.3390/ijerph18010271.
[23] T. P. Mashamba-Thompson and E. D. Crayton, "Blockchain and Artificial Intelligence Technology for Novel Coronavirus Disease 2019 Self-Testing," Diagnostics, vol. 10, p. 198, 2020, https://doi.org/10.3390/diagnostics10040198.
[24] D. Kuupiel, V. Bawontuo, and T. P. Mashamba-Thompson, "Improving the Accessibility and Efficiency of Point-of-Care Diagnostics Services in Low-and Middle-Income Countries: Lean and Agile Supply Chain Management," Diagnostics, vol. 7, p. 58, 2017.
[25] A. Albujeer and M. Khoshnevisan, "Metaverse and Oral Health Promotion," British Dental Journal, vol. 232, p. 587, 2022.
[26] Johns Hopkins Medicine, Johns Hopkins Performs Its First Augmented Reality Surgeries in Patients, 16 February 2021. Available at: https://www.hopkinsmedicine.org/news/articles/johns-hopkins-performs-its-first-augmented-reality-surgeries-in-patients (accessed on 22 July 2022).
[27] N. Kurian, J. M. Cherian, and K. G. Varghese, "Dentistry in the Metaverse," British Dental Journal, vol. 232, p. 191, 2022.
[28] K. I. Afrashtehfar and A. S. H. Abu-Fanas, "Metaverse, Crypto, and NFTs in Dentistry," Education Sciences, vol. 12, p. 538, 2022, https://doi.org/10.3390/educsci12080538.
[29] M. J. Sousa, F. Dal Mas, S. P. Gonçalves, and D. Calandra, "AI and Blockchain as New Triggers in the Education Arena," European Journal of Investigation in Health, Psychology and Education, vol. 12, pp. 445–447, 2022, https://doi.org/10.3390/ejihpe12040032.
[30] R. Q. Castro and M. Au-Yong-Oliveira, "Blockchain and Higher Education Diplomas," European Journal of Investigation in Health, Psychology and Education, vol. 11, pp. 154–167, 2021, https://doi.org/10.3390/ejihpe11010013.
[31] K. Nikolskaia, D. Snegireva, and A. Minbaleev, "Development of the Application for Diploma Authenticity Using the Blockchain Technology," Presented at the 2019 IEEE International Conference "Quality Management, Transport and Information Security, Information Technologies", IT and QM and IS, Sochi, Russia, 23–27 September 2019, 2019.
[32] D. Serranito, A. Vasconcelos, S. Guerreiro, and M. Correia, "Blockchain Ecosystem for Verifiable Qualifications," Presented at the 2nd Conference on Blockchain Research and Applications for Innovative Networks and Services, BRAINS, Paris, France, 28–30 September 2020, 2020.
[33] Y.-L. Chen, C.-C. Hsu, C.-Y. Lin, and H.-H. Hsu, "Robot-Assisted Language Learning: Integrating Artificial Intelligence and Virtual Reality into English Tour Guide Practice," Education Sciences, vol. 12, p. 437, 2022, https://doi.org/10.3390/educsci12070437.
[34] Y. Gao and L. Wu, "Efficiently Mastering the Game of NoGo with Deep Reinforcement Learning Supported by Domain Knowledge," Electronics, vol. 10, p. 1533, 2021, https://doi.org/10.3390/electronics10131533.
[35] Y. Tian, J. Ma, Q. Gong, S. Sengupta, Z. Chen, J. Pinkerton, and L. Zitnick, "Elf OpenGo: An Analysis and Open Reimplementation of AlphaZero," In Proceedings of the International Conference on Machine Learning, Long Beach, CA, USA, 10–15 June 2019, pp. 6244–6253, 2019.
[36] F. Han and M. Zhou, "Threat Matrix: A Fast Algorithm for Human–Machine Chinese Ludo Gaming," Electronics, vol. 11, p. 1699, 2022, https://doi.org/10.3390/electronics11111699.
[37] A. Predescu, D. Arsene, B. Pahonţu, M. Mocanu, and C. Chiru, "A Serious Gaming Approach for Crowdsensing in Urban Water Infrastructure with Blockchain Support," Applied Sciences, vol. 11, p. 1449, 2021, https://doi.org/10.3390/app11041449.
[38] S. Condino, M. Gesi, R. M. Viglialoro, M. Carbone, and G. Turini, "Serious Games and Mixed Reality Applications for Healthcare," Applied Sciences, vol. 12, p. 3644, 2022, https://doi.org/10.3390/app12073644.
[39] E. Choi, Y. Choi, and N. Park, "Development of Blockchain Learning Game-Themed Education Program Targeting Elementary Students Based on ASSURE Model," Sustainability, vol. 14, p. 3771, 2022, https://doi.org/10.3390/su14073771.

[40] I. Pittaras, N. Fotiou, V. A. Siris, and G. C. Polyzos, "Beacons and Blockchains in the Mobile Gaming Ecosystem: A Feasibility Analysis," Sensors, vol. 21, p. 862, 2021, https://doi.org/10.3390/s21030862.
[41] CryptoKitties Craze Slows Down Transactions on Ethereum. Available online: https://www.bbc.com/news/technology-42237162 (accessed on 22 August 2022).
[42] G. Yılmaz Benk, B. Badur, and S. Mardikyan, "A New 360° Framework to Predict Customer Lifetime Value for Multi-Category E-Commerce Companies Using a Multi-Output Deep Neural Network and Explainable Artificial Intelligence," Information, vol. 13, p. 373, 2022, https://doi.org/10.3390/info13080373.
[43] M. Qin, W. Zhu, S. Zhao, and Y. Zhao, "Is Artificial Intelligence Better than Manpower? The Effects of Different Types of Online Customer Services on Customer Purchase Intentions," Sustainability, vol. 14, pp. 3974, 2022, https://doi.org/10.3390/su14073974.
[44] D. O. Amofah and J. Chai, "Sustaining Consumer E-Commerce Adoption in Sub-Saharan Africa: Do Trust and Payment Method Matter?" Sustainability, vol. 14, p. 8466, 2022, https://doi.org/10.3390/su14148466.
[45] D. P. Sakas, N. T. Giannakopoulos, N. Kanellos, and C. Tryfonopoulos, "Digital Marketing Enhancement of Cryptocurrency Websites through Customer Innovative Data Process," Processes, vol. 10, p. 960, 2022, https://doi.org/10.3390/pr10050960.
[46] A. S. Bale, N. Ghorpade, M. F. Hashim, J. Vaishnav, and Z. Almaspoor, "A Comprehensive Study on Metaverse and its Impacts on Humans," Advances in Human-Computer Interaction, Hindawi, 2022.
[47] Google Trends. (n.d.). Google Trends. Retrieved July 3, 2022, from https://trends.google.com/trends/explore?q=metaverse&geo=IN.

Radhika, Kiran Bir Kaur, and Salil Bharany

10 Augmented Reality and Virtual Reality in disaster management systems: A review

Abstract: In this paper, we conducted a literature review to summarize the current state of AR and VR uses in real-world applications such as education, tourism, business marketing, storytelling, health and defense, retail and fashion, design and development in the context of IoT. This article examines AR/VR technologies used in disaster management and provides examples of how their combination delivers promising results in the applications mentioned above. It provides a thorough overview of the history of VR/AR/Mixed Reality technologies and their various applications. We discovered various benefits, possibilities, and challenges for disaster management, which we examined in detail. Additionally, we discussed the existing research gaps in XR (Extended Reality) technology for disaster management technology, which are necessary to assist disaster management effectively. These gaps were identified and explored to provide direction for future research. Moreover, we also present the development scope related to real-world industry applications and learning tools.

10.1 Introduction

Virtual Reality is described as the creation of a virtual world using computer technologies. After experiencing VR, the users will notice that it's a different perception than the one they have previously experienced. Virtual reality can be created artificially, for example by adding color to a picture, or it might be a snapshot of a real location that has been integrated with virtual reality software. The purpose of this technology varies depending on the application, but it mostly provides the users with information that they could not obtain from their senses alone [1]. Because AR/VR technologies have the potential to tackle a wide range of issues, prominent companies like Google, IBM, Sony, HP, and several universities have invested in their research. The terms "Virtual Reality" and "Augmented Reality" are used interchangeably. Virtual Reality and Augmented Reality can be applied to physics, chemistry, biology, mathematics, history, astronomy, medicine, and even music.

These large corporations are attempting to produce technological products that can accept any of these topics and, in turn, have an influence on the client life. The purpose of this article is to present

Radhika, Department of Computer Engineering & Technology, Guru Nanak Dev University and Global Group of Institutes Amritsar, 143001 Amritsar, India, e-mail: kumariradhika.178@gmail.com
Kiran Bir Kaur, Salil Bharany, Department of Computer Engineering & Technology, Guru Nanak Dev University, 143001 Amritsar, India, e-mails: kiran.dcse@gmail.com, salil.bharany@gmail.com

https://doi.org/10.1515/9783110785234-010

the concept of augmented and virtual reality, as well as the approaches that have been used to implement this plan of action. AR and VR are a technique that gradually overlaid 3D virtual elements onto the user's existing situation. We divide the particular requirements that must be met to provide the customer with the finest AR/VR experience possible in his environment. We also think about how AR/VR devices interface with certain environments and how clear they are. This overview's purpose is to present the up to the minute in augmented and virtual reality. The earliest half of this article is dedicated to the definitions of these distinct expressions, and the second section is dedicated to many applications of virtual and augmented reality systems in Disaster management systems [2] in different types of activities.

Disasters, whether they be caused by humans, natural forces, or a combination of both, are responsible for the massive loss of life and property. Appropriate disaster preparedness can have a significant impact on society. Disaster management is defined as "the body of policy and administrative choices, operational actions, players, and technology that correspond to all stages of catastrophes at all levels". Although technology is only one part of disaster management, it becomes clear as the conversation progresses that its position is inextricably linked to the other aspects. VR and AR technologies have uncovered several unique opportunities in the disaster management industry, including information exchange, rapid damage assessment, and reliable Building Information Modeling (BIM), among other things.

10.2 Definition of Augmented Reality, Virtual Reality and Mixed Reality

Recent advancements in Virtual Reality and Augmented Reality technology have enabled a new degree of human connection, resulting in a new method of interaction and knowledge sharing. Virtual Reality is a word that mentions to a set of technologies that allow a user to be immersed in a simulated 3D digital world while also generating interaction mechanisms that allow the user to interact with the environment and other users.

Virtual Reality In a virtual reality simulation, for example, participants can communicate, move from one location to another in a 3D digital environment, engage in manual tasks, operate machinery, drive vehicles, and interact with objects, events, and other participants in real-time. Each character can observe and respond to the actions of other characters, all while being immersed in a computer-generated environment that faithfully recreates the real world. The experts (simulation supervisors) can remotely monitor all activities undertaken by the managers running the virtual simulation, design localized events (such as damage to pipes and depletion of materials required for industrial processes), or introduce new challenges, like obstructions (e. g., tangled phone or electricity cords). Virtual Reality technologies have recently improved, allowing for the simulation of situation for which one wishes to train their employees in an increasingly realistic manner, making it a viable alternative to traditional training. By reducing the costs and development timeframes of actual exercises, simulated Virtual Reality environments can provide all the benefits of traditional training methods. Training that uses Virtual Reality technology to simulate disaster events can be scaled up or down depending on the organization that chooses to use Virtual Reality for specific purposes and training. For example, you can create a virtual scenario to provide immediate feedback to users about the input they receive (as in the case of organizing a triage of administrative procedures).

Augmented Reality [2] Augmented Reality, on the other hand, refers to a group of technologies that may "enhance" the real world by superimposing digital content, which the user perceives as a significant element of their viewpoint, onto actual reality. Augmented Reality, for example, can be used to learn more about a device, find a route, construct systems in specific environments with rooms and belongings, and examine the profile of a remote person as if they were in the same setting as the user. Virtual training systems enable and facilitate the collaborative training of geographically separated employees; for example, in the case of natural disasters such as earthquakes and hurricanes, virtual simulation can provide a consistent and synchronized training platform. Traditional teaching methods such as lectures are no longer as effective. For example, slide presentations can be easily integrated into a virtual simulation system and transferred to a new format. While the benefits of virtual simulation systems are clear in terms of visual stimuli, it is also important to emphasize the benefits of acoustic stimuli that can be introduced into a virtual environment. One of the variants of the Virtual Environment (VE) or Virtual Reality (VR) is Augmented Reality (AR). The distinction between Augmented Reality and the more commonly used Virtual Reality is that Virtual Reality aims to immerse the user in a virtual world. Augmented Reality, on the other hand, brings digital or computer-generated information into the real world. Images, music, video, or haptic (tactile) senses that overlap in the real world can all be used to convey information. As a result, Augmented Reality allows the user to observe the natural world while also projecting virtual objects onto it. The appropriate implementation of Augmented Reality necessitates the use of specific components. The components required for Augmented Reality are as follows [3]:

- Computer or mobile device
- Monitor or another type of display device
- Camera
- Tracking and sensing systems (GPS, compass, accelerometer), which are found in almost all smart devices
- Network architecture
- Tags, which are tangible objects or locations that connect real and dynamic environments. In this step, the computer selects the area where the digital information will be displayed.

- Software or application that runs on a single computer
- Web services
- Content server

Possibilities for incorporating Augmented Reality technology into crisis management The potential for augmenting the real world with Augmented Reality is boundless. In the workplace, this technology can be found in the form of a teaching backdrop or approach for employees, smart device games, or instructional materials. Population protection and crisis management are critical parts of human existence. When analyzing the relationship between Augmented Reality and these fields, it's possible that acquiring knowledge about how to operate in these areas could be simplified and more efficient. Not only are there rules and regulations for population protection and crisis management, but there are also organizations that represent them. The fire department, the police force, and the medical rescue service are all part of an integrated rescue system [4]. In the future, Augmented Reality may play a part in the critical performance of their responsibilities or exercises.

"Improving Our View of the World: Police and Augmented Reality Technology" explores "probable augmented reality uses for New York City police" [4]. Potential uses of this technology include:
- Implementation of intelligent real-time crime reporting
- Collection of face scans, voiceprints, and other biometric data on known offenders
- Scalable, three-dimensional maps providing detailed building and sewer floor plans
- Improved team cohesion and coordination with the intervention commander
- Modification of gunshot sound effects to enhance attention
- Thermal and infrared imaging, as well as optical zoom
- Visualization of blood samples, bloodstains, and other sensor-detectable forensic data at crime scenes using Augmented Reality video, audio, and sensing equipment
- Collaboration between robots, unmanned aerial vehicles (UAVs), and law enforcement
- Viewing 3D projections of location, activity, and status information
- Coordination of widely dispersed units [1]

Mixed Reality (MR) Mixed reality is a composite reality in which the physical and digital worlds intersect to create new environments and representations that coexist and interact in real-time with both tangible and digital objects. MR is a combination of Augmented Reality and Augmented Virtuality that can be found in both the physical and virtual worlds. Holography [4] is a term used to describe MR's ability to integrate digitally created objects into the real environment. In Mixed Reality, real and virtual content coexist and interact in real-time. It uses a combination of Augmented Reality and Virtual Reality. MR is more than a substitute for Augmented Reality or Virtual Reality; it is a unique experience that enriches users' understanding of the interactions in both the real and virtual worlds. Flexibility, immersion, engagement, cohabitation, and enrichment are all required in a mixed reality experience. This is achieved through the utilization of Augmented Reality and Virtual Reality (AR/VR) technologies. As a result, a Virtual Reality (VR) experience generates a Mixed Reality environment in which individuals feel immersed, and their perception of the real world is enhanced [5].

The Reality-Virtuality continuum This term is used to describe the relationship between reality and Virtual Reality. In this context, the continuum comprises all possible variations and combinations of real and virtual objects, ranging from fully real to completely virtual.

Due to the similar underlying technologies, which provide an enhanced experience with full immersion, users have confused AR with VR, mistaking the two technologies for one another. The Reality/Virtuality Continuum, developed by Milgram in 1994, helps clarify this confusion [6]. Milgram defined Mixed Reality by introducing the reality-virtuality continuum and identifying a range of technologically changed patterns of reality that match today's Augmented and Virtual Reality technologies.

Augmented Virtuality (AV) The ability to interact with a virtual representation derived from the real world is known as Augmented Virtuality (AV). AV is a subcategory of MR that integrates real objects into a virtual environment. This is usually achieved by streaming video from physical space (via a webcam, etc.) or 3D scanning a physical object, video, or object and integrating it into a virtual object, where the virtual world appears somewhat like reality and maintains flexibility in the virtual world [9]. For example, an aircraft maintenance engineer visualizes a real-time model of an aircraft engine in flight so that real-world elements appear on an animated screen.

Comparison Augmented Reality, Virtual Reality, and Mixed Reality can all be used to achieve the objectives mentioned above, but our research suggests that Augmented Reality is the best option for facilitating an exhibition. Virtual Reality, on the other hand, seems to be suitable for virtual museums, while Mixed Reality seems to be more practical in both indoor and outdoor reconstruction applications. Here are some simple work definitions:

Augmented Reality (AR) By overlaying two-dimensional virtual information onto our view of the real world, Augmented Reality aims to enhance our perception and understanding of the real world.

Virtual Reality (VR) Virtual Reality aims to improve our interaction and immersion in a computer-generated environment without allowing us to be involved or view the real world.

Augmented Virtuality (AV) The goal of Augmented Virtuality is to enrich the virtual environment with real-life information.

Mixed Reality (MR) The goal of Mixed Reality is to combine real and virtual worlds.

In recent decades, numerous new technologies have been proposed to mitigate the impact of disasters on human and built environments. Among them, Virtual Reality (VR) and Augmented Reality (AR) represent two innovative technologies that have demonstrated their potential in some cases to support the design of safer build environments and enhance safety training [8]. Global research opportunities in this area are abundant.

AR/VR technologies, which are still in their early stages of development, have a myriad of possible applications. The following are some examples of research directions specifically focused on emergency operator preparation:

- VR: In a digital context, create new modalities of mobility.
- VR: Design an interactive virtual tour for educational purposes using a "gamebook" approach.
- AR: Develop Augmented Reality information to enable efficient physical and digital interactions in performing manual tasks (e. g. connecting up a hose, selecting the appropriate tool, etc.) and joining the SAFE network.
- VR/AR: Create new tools and methods tailored to meet the most frequent demands in the digital environment (e. g., companion tracking, communication, writing, reporting, geo-referencing, object management, inventory management, etc.).
- VR/AR: Design new algorithms for candidate evaluation (scoring, computation, reporting).
- VR/AR: Develop of technology platforms that contribute to training and optimize training regimens to maximize the efficiency of VR and AR components.

10.3 AR, VR, and MR implementation issues

Many breakthroughs are being made to advance AR-VR technology, but there are still several concerns that have to be addressed, which act as major roadblocks to AR and VR's expansion in the market and among the general public. The following are some of the critical challenges that must be addressed:

1. Dedicated hardware requirement: The most challenging aspect is the demand for dedicated hardware in the initial setup. Virtual Reality applications require a specific space and environment.
2. Need for a less expensive technology: Customers are not yet ready for AR-VR products because they come at a high price. As a result, comprehensive solutions are needed that make use of less expensive and more efficient hardware, reducing the long-term cost of AR-VR items.
3. Lack of real use cases: Even if the cost issue is resolved, AR/VR still face significant challenges in the form of creative and unique content. Current R&D efforts primarily focus on the gaming and entertainment industry. The material produced should reflect the consumer's point of view. Entertainment and gaming are currently accessible, but these technologies have yet to discover applications that will make them indispensable to consumers and businesses.

4. Issues with mobility and miniaturization: Mobility is one of the most fundamental challenges with VR experiences. In some VR products, a tangle of cables connected to HMDs (Head Mounted Displays) or other wearable equipment obstructs unrestricted mobility. Advances in VR products should be small, light, compact, and portable, allowing customers to have a cordless or wireless experience with ease.
5. Low concern for security: Cyber security is another significant concern with AR-VR technology that has yet to be effectively addressed. There's a risk that virtual environments will be hacked, with attackers gaining access to, modifying, or changing them, as well as destroying them. As a result, security measures will need to be developed in the future since AR-VR components connect directly to phones, desktops, and laptops.

10.4 Advantages vs. constraints

Virtual Reality-based training systems provide several advantages over traditional training. It is easy to depict real-life settings that incorporate other end-users from an environmental perspective; moreover, user inputs may be controlled to offer real-time feedback. By populating the simulated avatar, it is possible to deliver immediate feedback to the learner (presentation of users in the dynamic environment). Various avatars can be given behavior modules that regulate the decisions made by the person being trained. Virtual Reality systems also allow the adjustment of the difficulty level depending on the preparedness of the individuals who need to be trained. Objects in the virtual world, as well as other training participants, can be interacted with during the session. Text and/or voice communication can be used to communicate with real and virtual individuals, allowing for the modeling of diverse cultural and socioeconomic circumstances. One can freeze (pause) the virtual simulation to discuss how instructors and trainees can effectively deal with unexpected events. If desired, the teacher can make some adjustments and repeat the activity [6].

Buildings, greenery, individuals, and sounds are just a few of the elements that can be easily included in the environment. Although real-life large-scale training typically takes more time and money, virtual proposals such as the development of safe zones within densely populated urban areas can be easily generated. These latter features are one of the most significant advantages of virtual practical training systems, as real-world exercises can be prohibitively expensive for most governments. Virtual Reality offers a viable alternative for various components of the assignment. After mastering all of the initial parts, operators can immediately begin working on the additional factors of difficulty (heat, weight, odors, etc.) when they transition to real-life training [7]. A virtual workout, on the other hand, can be conducted multiple times at a minimal cost. As a consequence, operators improve their mobility on some of the more challenging components of the assignment. Additionally, the virtual scenario's versatility allows for the simulation of multiple settings, enabling the evaluation of alternative outcomes under various simulation scenarios. Finally, because virtual simulations are digital, data can be preserved for subsequent analysis and review. This can be useful in emergencies, as it provides a better understanding of behavioral patterns and helps develop more effective training processes. It's essential to keep in mind that Virtual Reality should be viewed as a supplement to traditional training methods rather than a replacement.

Furthermore, AR/VR technologies have the potential to dramatically improve the efficiency of communication and training processes, particularly through the techniques outlined below [7]:

- **Cost savings:** Digital information improves reproducibility, allowing for multiple repetitions of people and vehicle movements and complex training procedures (such as emergency training) when required.
- **Applicability to all sectors and situations:** Digital scenarios can simulate any type of environment, object, or effect that can be tailored to specific context needs, training types, or user learning preferences, with options to insert difficulty elements. Test candidate reaction times for unexpected events in real-time.
- **Overcoming geographical boundaries:** People can engage even when they are geographically separated, experiencing and participating in the same environment, due to the ability to access AR/VR content remotely. This capability offers up numerous possibilities. For example, Italian civil defense operators can conduct regular training and test new intervention procedures in civil defense in other countries.
- **Multiple users connected simultaneously:** When AR/VR technology is integrated into the internet, it is possible to create multiplayer scenarios where several people interact in an immersive digital world. This component can generate new logic for training process planning ("field training" in a digital environment, the possibility of training multiple operators simultaneously, and strengthening the team's coordination mechanism) [6].
- **"Errors" are automatically assessed:** Each action may be tracked in real-time, as each user's movements in space can be recorded. This allows the development of automatic error evaluation algorithms that provide suggestions on how to improve certain physical work activities or approaches over time, both individually or as a part of a team.

However, these technologies are not without drawbacks. In the digital realm, motion sickness (vertigo) difficulties are influenced by visual quality, frame rate, and forced movement mechanisms (shifting, arm swing, legs wing, free movement, and so on). Other impressions that are part of the reconstructed event are impossible to repeat (heat, scent, etc.).

It's essential to understand that a virtual environment cannot fully replace real-world experiences; it can only simulate them to a limited extent, covering approximately 30–40 % of the aspects (environments, actions, coordination between people). However, a virtual workout, has the advantage of being conducted multiple times at a minimal cost. As a result, operators can improve their skills on some of the more challenging tasks.

10.5 Applications of AR/VR in different areas

Gaming Augment Reality can open up many doors in the game production business, allowing users to play and interact with virtual objects in the real world, creating an immersive experience where the user can engage with both the real-world surroundings and the augmented virtual elements. The ability to generate landscapes depending on the user's surroundings in real-life may be included in future games. Virtual Reality (VR) gaming, on the other hand, has advanced the user's gaming experience to the point where the VR environment is perceived as the real world. Due to the 3D VR environment and developments in the control system, users can experience a heightened sense of immersion and engagement within the dynamic virtual environment, allowing them to fully participate in various actions and focus on the sensations presented. The success of games like "Pokemon Go" has showcased the true potential of AR in the gaming entertainment industry.

Education Students will be able to understand study topics more dynamically using vibrant Expanded Reality materials in classroom lectures or practical lab sessions, which will encourage them to learn by delivering more information on the subjects in numerous manners. Teachers and students can be consistently engaged in interactive activities involving augmented content throughout class sessions. Virtual Reality may be used as an interactive medium for experimentation, recreating real-world environments, and other comparable activities. In a virtual context, it has the power to bring learning to life. Students are more likely to feel a personal connection to the educational subject matter. They can now explore intellectual concepts that were previously inaccessible, enhancing their learning experience.

Fashion applications AR can offer a way for people to try on items without needing a fitting room, allowing them to virtually wear and try on their favorite fabrics, shoes, purses, and jewelry. These materials may be virtually placed on the customer's body, providing information about the item's fit and appearance based on their body measurements. This has the potential to change people's shopping habits when it comes to fashionable items. Virtual reality, on the other hand, has the potential to create worlds in which new products and fashion trends may be developed and tested and showcased to obtain feedback from customers by presenting new fashion materials. This may be used to create new things, to recreate and promote fashion styles including garment quality creation, jewelry fine art design, and many more.

Healthcare In medical procedures, Augmented Reality can be used to visualize hidden body parts and simulate the 3D object of the body part being operated on, particularly for medical training purposes. AR's real-time interactive data can be used to visualize the human anatomical structure to practice and improve surgical abilities. Instead of only providing long descriptive information written over pill bottles, pharmaceutical businesses can use AR technologies to demonstrate how a particular medicine will work in 3D view. In contrast, Virtual Reality (VR) can be utilized for virtual surgical training, with no risk of practicing on real patients. It can also be employed in meditation treatment methods, as well as therapies and treatments for phobias, anxiety, and pain management in patients.

Military Armed Forces AR can be used to provide soldiers with real-time weapons information. This data can be used to find out how, when, and under what circumstances a weapon can be used. The weapon-mounted system can be demonstrated to soldiers during training so that they can interact with the weapons. Virtual reality (VR) can be employed to develop various virtual training environments for soldiers, simulating battle scenarios with advanced control systems that can be used to train soldiers in making physical and psychological decisions during warfare. To enhance simulations, Augmented Reality technology is utilized in operating drones during military training. AR/VR/MR applications, on the other hand, are not limited to the mentioned sectors; instead, these technologies have the potential to revolutionize various experiences, including but not limited to training, and marketing experiences.

This paper brings together a collection of works that propose AR or VR applications to improve catastrophe preparedness and safety in the built environment. The works were grouped into three categories based on their contributions to the field:
(a) Improving the design of constructed environments for safety
(b) Investigating how individuals behave in the aftermath of a calamity
(c) Training people to respond to crises.

It's worth noting that several of the works have different goals and have appeared in several collection papers.

10.6 Empirical Analysis

For this study, a review process was employed to track and synthesize the literature. The scope of this article is two-fold:
- Disaster management
- XR technology

After a comprehensive investigation of structured review methods, the research [9] was determined to be valuable for engineering and disaster management within Tran's dis-

Table 10.1: Summary of PICO.

Population	**XR stands for Tragedy recovery mechanics.**
Intervention	Research articles upon catastrophe management technologies from XR.
Comparison	Conventional technique vs. XR
Outcome	XR research's scope and usefulness in disaster management

cipline area. The document was used to perform the analysis and provide the outcomes which included:

1) Establishing the opportunity
2) Tracking resources
3) Analysis
4) Synthesizing
   1. Defining the study's vision and research: Use the PICO framework (Table 10.1) to define the scope of your study (PICO) by constructing populations, interventions, comparisons, and outcomes. To assess the effectiveness of the technology, compare research articles on XR applications with regular disaster management tactics. The entire study was constructed using various mixed design approaches, as the potential results of the study require both quantitative and qualitative research. Qualitative output has higher priority than quantitative results, so built-in research focuses on portraying application areas that have higher priority than effectiveness and limitation. The PICO technique was used. Additionally, the results were delivered on time.
   2. Acceptance criteria: The primary objective of the research report was to explore new research trends in application development for XR technology in support of emergency and disaster risk management. Only articles written in English were subject to review. The search for studies was conducted using specific keywords that included the terms "disaster" and "XR technology."
   3. Locating and organizing sources: In order to discover progressive research papers, which are often presented at conferences and published in pivotal research journals, the tracking had to be chosen. Despite the high number of documents provided by Google Scholar, the great majority of them were only related to only one domain. Although the authors needed to use a large number of keywords, the database retrieved publications from certain sources, significantly contributing to tracking down original research efforts on this multidisciplinary topic [10].
   4. Tracking source settings: These settings were updated according to the eligibility criteria. The search for articles was then conducted using the aforementioned search method. Each article was examined individually based on the title displayed in the tracking source results. Due to the interdisciplinary nature of the review, the screening process involved examining the paper titles

for XR-related terms such as "augmented reality," "virtual reality," and "mixed reality," as well as disaster management-related terms like "flood," "response," and "emergency," or "Disaster," Some titles used terms such as "wearable," "simulator," "system," and disaster-related terms. However, they were all based on 3D computer graphics. This stage was important for screening large numbers of documents on Google Scholar. Next, the abstracts and introductions of only relevant papers were reviewed in more detail. This stage was essential for eliminating patent documents. Finally, useful information from the selected studies was edited and documented under the titles Keyword, Article Title, Year of Publication, Track Source, Work Description, Disaster Focus, Findings, and Research Issues. During this time, certain studies using XR for crisis management did not necessarily develop an XR system; instead, the technology, was only used to conduct behavioral research. Such publications were manually identified and excluded from the review.

10.7 Methods

AR and VR have been used in various ways to reduce the impact of disasters on humans and the environment they inhabit. Several reviews have been published to date, highlighting the potential of these technologies for specific purposes (such as education and human behavior research) or specific disasters (such as fire research) [11]. Instead, this section aims to explain the different areas in which these technologies are used. Some of the work included in the previous review has been integrated and additional work has been added. Using Google Scholar, the following keywords were identified: Virtual Reality, Augmented Reality, Evacuation Virtual Reality, Augmented Reality, and Evacuation Training. The works selected for this paper met both the following criteria:
(a) An AR or VR application for building evacuation was proposed;
(b) An AR or VR application was tested through experiments.

Disaster Preparedness Training AR and VR offer significant potential for education across different areas. This also applies to the field of disaster protection research, where various VR and AR applications are being developed to prepare for disasters such as earthquakes, tsunamis, tornadoes, plane crashes, and building fires. The primary objective of the research review was to investigate new research trends in application development for XR technology that supports emergency and disaster risk management. Articles published in journals, conferences, and book chapters were reviewed. Only works in English were considered for evaluation. Papers were found with specific keywords including the terms "disaster" and "XR technology."

Extended Reality (XR), which encompasses Augmented Reality (AR), Virtual Reality (VR), and Mixed Reality (MR), is a collection of computer graphics programs that allow users to work with computer-generated images in the real and virtual worlds. Augmented Reality scenarios consist of a real environment with non-dominant virtual elements that react to users and scenes [12]. A VR scene consists of a completely virtual environment in which virtual elements obscure the physical environment and mimic physical objects. In contrast, AR scenes consist of real-world environments where integrated but non-dominant virtual objects are directed at the user or the scene can react.

Table 10.2 illustrates that there is a lot of variation in hardware settings for both VR and AR investigations and that some studies have looked into using several setups. The

Table 10.2: Hardware setup and training goals of various disaster related research papers.

Reference	Hardware setup	Type of Disaster	Training Goals
Feng, González, Mutch, et al., 2020	VR-HMD	Earthquake	Earthquake Preparedness
Mitsuhara & Shishibori, 2020	VR-HMD AR-VST	Tornado	Tornado awareness
Li, Liang, Quigley, Zhao, & Yu, 2017	VR-HMD	Earthquake	Drop to cover and hold
Ruggiero Lovreglio, Duan, Rahout, Phipps, & Nilsson, 2020	VR-HMD	Building Fire	Use of Fire extinguishers
Månsson & Ronchi, 2018	VR-HMD	Building Fire	Use of Fire extinguishers
Feng et al., 2019	VR-HMD	Earthquake	Earthquake preparedness
Bright & Chittaro, 2016	VR-HMD	Aircraft accident	Location of emergency exits
Chittaro & Buttussi, 2015	VR-HMD	Aircraft accident	Brace position and evacuation procedures.
Smith & Ericson, 2009	VR-CAVE	Building Fire	Fire evacuation procedures
Kinateder et al., 2013	VR-CAVE	Tunnel Fire	Fire safety behaviors
Farra et al., 2019	VR-Non immersive VR-HMD	Building Fire	Evacuation of neonates
López, Plá, Méndez, & Gervás, 2010	AR-VST	Tsunami and earthquake	Evacuation procedures
Kawai, Mitsuhara, & Shishibori, 2016	AR-VST	Tsunami and earthquake	Evacuation procedure
Mitsuhara, Shishibori, Kawai, & Iguchi, 2016	AR-VST	Tsunami	Evacuation process
Mitsuhara, Iguchi, & Shishibori, 2017	AR-VST AR-OST	Earthquake	Earthquake preparedness
Mitsuhara, Iwaka, et al., 2017	AS-VST	Tsunami and earthquake	Location of emergency exits
Sharma, Bodempudi, Scribner, Grynovicki, & Grazaitis, 2019	AR-VST AR-OST	Building fire	Location of emergency exits

majority of the studies listed have been used to educate individuals on a variety of skills for dealing with disasters, while others have focused on specific safety activities such as the use of fire extinguishers.

10.8 Conclusion

This paper provides an overview of existing Augmented Reality (AR) and Virtual Reality (VR) hardware setups that can be used for disaster-related research. It also includes an overview of applications developed in recent decades to enhance human safety in the face of calamities. The analysis included 64 studies that explored the potential of AR and VR for three main purposes:
(a) Improving the safety of built environments
(b) Studying human behavior
(c) Training people

This work demonstrates that researchers have a vast array of hardware setup options to conduct research on human behavior in disasters. The selection of the optimal setup depends on several factors such as research budget and research goals. For instance, some VR and AR setups, such as VR-CAVE and AR-OST, remain expensive compared to alternative VR and AR hardware setups. When choosing a hardware setup, it is essential to consider that the options listed in Section 10.2 can generate different levels of immersion for users and interactions with digital elements. Tables 10.1 and 10.2 identify only a few studies that have compared the research output generated while using different VR and AR setups to investigate the same research question, for instance [13] for VR setups comparisons for AR setups comparisons. Although these works provide preliminary results on how different setups works and their advantages and limitations, it is necessary to gather more evidence with future studies. To date, only one study has tested both VR and AR solutions to enhance disaster awareness.

According to the review, 23 studies have utilized Virtual Reality and Augmented Reality to improve the safety design of built environments. This was achieved by either asking participants to compare and score different layouts or by observing how individuals responded when exposed to different layouts. The research shows that Virtual Reality studies have only focused on fire disasters, and further research is needed to determine if similar approaches can be used to investigate safety solutions for other types of disasters. Studies suggest that using a holographic guiding system to help evacuees choose the best path to escape is advantageous. However, only one research has investigated the potential of AR-OST devices. This is most likely due to the scarcity of these devices in recent years, as well as their high costs. Nonetheless, with the recent availability of new AR-OST devices, more applications and studies using AR-OST are expected in the coming years [14].

According to the study, several VR systems have been widely used to investigate how individuals react to disasters such as building fires, earthquakes, and wildfires. However, one of the major problems is the ecological validity of the data obtained through Virtual Reality (i. e., if humans behave similarly in Virtual Reality and actual disasters) [15]. Several research papers have been conducted to address this fundamental issue. To provide quantitative assessment of the ecological validity of VR research, further studies comparing actual and virtual studies are needed. In addition, according to the review no AR study has been conducted to explore human behavior in disaster situations.. As mentioned earlier, AR research has a wide range of possibilities.

Finally, this study illustrates that VR and AR have been used in numerous studies on human behavior in disasters. According to the findings, the majority of these studies have concentrated on fire disasters, especially in terms of safety design and behavioral investigations. The review suggests that further research is required to analyze the benefits and drawbacks of different VR and AR hardware configurations and to validate the results obtained using these new emerging technologies.

10.9 Future Directions

Computer simulation modeling Creating a simulation requires substantial computational power. We recommend using stable hardware when constructing and running your simulations. Otherwise, it can cause buffering and slowdown of image transitions as well as position tracking in VR, AR, and MR systems. At the most basic level, current hardware designed to accelerate processing on the GPU creates realistic visualizations that can enhance the ecological validity of the simulation. According to [16], adding location-based technologies to disaster management systems such as building information modeling (BIM), geographic information systems (GIS), and RFID (Radiofrequency Identification) can significantly improve disaster management systems [17].

Research has shown that many research papers on creating unique simulations have not been adequately tested and validated. This indicates that the use of XR in disaster management is still in its infancy and needs further investigation. During the investigation, several incident command response systems were discovered. XRs (Common Operational Pictures), cloud computing (efficient communications and information storage), and geo networks are among the key system integration components (deployment and deployment of an array of sensors to collect disaster data) bottom. Several drone-related attempts were discovered during the audit. Drones, on the other hand, were not used in the search and rescue command system. Integrating drones with incident management systems could improve resilience.

In the field of 3D reconstruction, uncalibrated photos caused stretch and crash issues for 3D models. Further research is needed to develop a method for properly aligning the photo to the 3D geometry so that simulation quality is not affected by the image quality. It is also a desirable characteristic to make the simulation as realistic as possible. On the other hand, individuals involved in very severe disaster simulations can experience mental distress. Balancing realism and mental safety of disaster-related simulations could be a research area of interest in the study of XR and safety science [18].

Finally, general emergencies are divided into two categories in the realm of disaster management: routine crises and crisis emergencies [17]. The literature on disaster management provides a comprehensive explanation of each disaster, as well as the essential variables that contribute to an effective response. The development of these simulations based on principles from the literature on disaster management can provide learners with the skills they need to adapt to unexpected catastrophic situations.

Interaction techniques With the development of new interaction tools and methodologies, researchers and scientists can now interact with a range of scenarios in an immersive virtual world. According to [18], eye-tracking helps us to identify and analyze more complex and dynamic architectural environments. Newer devices, such as haptics, can enhance interaction and make an XR system more immersive. Similarly, AR technology is expected to be adaptable to various settings. According to future research on performance adaption to all weather conditions, it could boost the technology's usage. Computer vision improvements for AR research might be beneficial. Such applications could benefit from remote sensing for XR. For AR visualizations, a suitable solution for augmenting the adaptive latency compensation algorithm with image processing methods is also necessary. On the other hand, latency issues have a direct impact on the interaction aspect [17].

Lastly, the development of evacuation assistance systems requires a deep understanding of human navigational behavior. Researchers should investigate how difficult basic cognitive processes impact contiguous investigation mode in fire crises, according to [19]. They observed that it might be used to develop better behavior involvement methods and tools to assist with effective building evacuation in the case of a fire.

Training Successful training simulations require realistic training environments with hardware-accelerated code and sufficient sensors. According to the author, marker-based AR training should be avoided or used in simulator training with markers that blend into the simulation. This is because virtual elements are brought into the real world using markers, and trainees can notice markers and paper associated with moving objects, degrading the quality of the simulation. If marker-based AR is necessary, creative markers are highly recommended, such as blood images as markers for placing virtual human bodies. Intelligent algorithms can be used to recognize objects in the input image and determine where to overlay virtual objects as a better alternative. On the other hand, simulators for disaster planning and response developed according to user-centered design principles are promising. Using a user-centered model not only provides an easy-to-use product [14] in a short amount of time but also improves the user's perception and experience of the virtual world. In addition, the contextual cognitive framework is also highly effective from an educational perspective in the virtual world [16]. Some proposed system design methods for developing effective XR-based training simulations requirements analysis, interface prototyping, comprehensive system prototyping, and cross-functional system prototyping [17]. As XR technology continues to expand rapidly into other fields, educational research on how XRs can be designed to enhance student learning is becoming also a growing area of professional interest. A VR lesson model for bridge construction using the cantilever and stepwise stacking methods is described by [19]. These models can be used to develop effective training simulators for construction workers and increase workplace safety.

Public awareness Only a few publications were found that used XR technology to inform the general public about disasters. Further research can be conducted to develop tools and techniques for educating the general public about disaster risk. The SG program presented in [20] claims that it can be tailored to local environmental issues by creating a custom sequence of interactions and challenges currently being developed for Manchester and the North-West of the United Kingdom. Collective consciousness is another area of research and should not be incorporated into training. They expressed concerns about the use of virtual reality and argued that it could be potentially used to mislead people in disaster situation into fleeing. If this is not predicted before implementation, technical validation is needed to avoid widespread disasters.

Infrastructure assessment and reconnaissance Transferring hardware capabilities to software can be beneficial for remote assessment and reconnaissance. Mobile AR with cloud processing is a popular option, as it documents and critically synthesizes the corpus to provide an assessment of current trends and future proposals. The use of XR as a disaster management technique is widespread in the research community. Despite its binding content [21], the information medium is crucial for the proper transmission of information. Systems based on XR technology not only excel at retaining spatial working memory and improving knowledge but also reduce the time and number of errors spent on tasks, compared to the more practical and traditional methods. XR systems are widely used in crisis management applications such as computer simulation modeling, interactive techniques, training, infrastructure assessment and reconnaissance, and public awareness. We briefly considered the priorities, outcomes, and limitations of all studies. However, more research is needed to address health concerns related to VR testing and validation, computing and sensor hardware integration, acquisition, and mapping automation, to better service the science of disaster management using XR technology. The results of this study are particularly relevant to research and industrial organizations seeking to create successful and cost-effective XR-based disaster management systems.

Bibliography

[1] P. Hodgson, V. W. Lee, J. C. Chan, A. Fong, C. S. Tang, L. Chan, and C. Wong, "Immersive Virtual Reality (IVR) in Higher Education: Development and Implementation," In Augmented Reality and Virtual Reality: The Power of AR and VR for Business, pp. 161–173, Springer, Cham, 2019.

[2] M. Mendes, J. Almeida, H. Mohamed, and R. Giot, "Projected Augmented Reality Intelligent Model of a City Area with Path Optimization," Algorithms, vol. 12, no. 7, p. 140, 2019, https://doi.org/10.3390/a12070140.

[3] S. Kumari and N. Polke, "Implementation Issues of Augmented Reality and Virtual Reality: A Survey," vol. 26, Springer International Publishing, 2019.

[4] V. Kohli, U. Tripathi, V. Chamola, B. K. Rout, and S. S. Kanhere, "A Review on Virtual Reality and Augmented Reality Use-Cases of Brain-Computer Interface Based Applications for Smart Cities," Microprocessors and Microsystems, vol. 88, 104392, 2022.

[5] D. Schmalstieg and T. Hollerer, "Augmented Reality: Principles and Practice," Addison-Wesley, Boston, MA, 2017.

[6] C. Z. Li, J. Hong, F. Xue, G. Q. Shen, X. Xu, and M. K. Mok, "Schedule Risks in Prefabrication Housing Production in Hong Kong: A Social Network Analysis," Journal of Cleaner Production, vol. 134, no. Part B, pp. 482–494, 2016.

[7] X. Liu, X. Wang, G. Wright, J. Cheng, X. Li, and R. Liu, "A State-of-the-Art Review on the Integration of Building Information Modeling (BIM) and Geographic Information System (GIS)," ISPRS International Journal of Geo-Information, vol. 6, no. 2, p. 53, 2017.

[8] D. Azougagh, Z. Rabbani, H. Bahatti, A. Rabbani, and O. Bouattane, "Auto-Fit Multiple Movable Projections Into One Display Screen," In Int. Conf. Wirel. Technol. Embed. Intell. Syst. WITS 2019, pp. 0–3, 2019. https://doi.org/10.1109/WITS.2019.8723667.

[9] Z. Rebbani, D. Azougagh, A. Rebbani, H. Bahatti, and O. Bouattane, "Auto Guiding a Mobile Projector," In ACM Int. Conf. Proceeding Ser., 2018, https://doi.org/10.1145/3289402.3289503.

[10] N. Menck, C. Weidig, and J. C. Aurich, "Virtual Reality as a Collaboration Tool for Factory Planning Based on Scenario Technique," Procedia CIRP, vol. 7, no. December 2013, pp. 133–138, 2013, https://doi.org/10.1016/j.procir.2013.05.023.

[11] J. Grubert, Y. Itoh, K. Moser, and J. E. Swan, "A Survey of Calibration Methods for Optical See-Through Head-Mounted Displays," IEEE Transactions on Visualization and Computer Graphics, vol. 24, no. 9, pp. 2649–2662, 2018, https://doi.org/10.1109/TVCG.2017.2754257.

[12] A. R. Caballero and J. D. Niguidula. Disaster Risk Management and Emergency Preparedness: A Case-Driven Training Simulation Using Immersive Virtual Reality. In Proceedings of the 4th International Conference on Human-Computer Interaction and User Experience in Indonesia, CHIuXiD'18, pp. 31–37, 2018, March.

[13] V. Salehi, H. Zarei, G. A. Shirali, and K. Hajizadeh. "An Entropy-Based TOPSIS Approach for Analyzing and Assessing Crisis Management Systems in Petrochemical Industries," Journal of Loss Prevention in the Process Industries, vol. 67, 104241, 2020.

[14] D. Freeman, S. Reeve, A. Robinson, et al., "Virtual Reality in the Assessment, Understanding, and Treatment of Mental Health Disorders," Psychological Medicine, vol. 47, no. 14, pp. 2393–2400, 2017.

[15] D. Ma and H. G. Kim, "Shape of Light: Interactive Analysis of Digital Media Art Based on Processing," Techart Journal of Arts and Imaging Science, vol. 7, no. 4, pp. 23–29, 2020.

[16] K. Tarutani, H. Takaki, M. Igeta, et al., "Development and Accuracy Evaluation of Augmented Reality-Based Patient Positioning System in Radiotherapy: A Phantom Study," In Vivo, vol. 35, no. 4, pp. 2081–2087, 2021.

[17] A. Alper, E. E. Zta, H. Atun, D. Çinar, and M. Moyenga. "A systematic literature review towards the research of game-based learning with augmented reality," International Journal of Technology in Education and Science, vol. 5, no. 2, pp. 224–244, 2021.

[18] S. Park, S. H. Park, L. W. Park, S. Park, S. Lee, T. Lee, et al., "Design and Implementation of a Smart IoT Based Building and Town Disaster Management System in Smart City Infrastructure," Applied Sciences, vol. 8, no. 11, p. 2239, 2018.

[19] A. Palance and Z. Turan, "How Does the Use of the Augmented Reality Technology in Mathematics Education Affect Learning Processes: A Systematic Review," Uluslararası Eğitim Programları ve Ogretim Calısmaları Dergisi, vol. 11, no. 1, pp. 89–110, 2021.

[20] J. Lin, L. Cao, and N. Li, "Assessing the Influence of Repeated Exposures and Mental Stress on Human Wayfinding Performance in Indoor Environments Using Virtual Reality Technology," Advanced Engineering Informatics, vol. 39, pp. 53–61, 2019, https://doi.org/10.1016/j.aei.2018.11.007.

[21] Zhang, J. Sun, and J. Liu, "Diagnosis of Chronic Kidney Disease by Three-Dimensional Contrast-Enhanced Ultrasound Combined With Augmented Reality Medical Technology," Journal of Healthcare Engineering, vol. 2021, no. 3, Article ID 5542822, 12 pages, 2021.

M. S. Sadiq, I. P. Singh, M. M. Ahmad, and M. Babawachiko

11 Virtual Reality convergence for Internet of Forest Things

Abstract: The promotion of sustainable forest management of natural resources is hampered by environmental sustainability and climate change. In recent years, the Internet of Things (IoT) technology has advanced swiftly and been successfully used in various areas. The IoT application context now includes a wider range of applications. The significant potential of augmented, virtual, and hybrid realities (AR/VR/MR) technology stems from the fascinating and satisfying experiences they can provide.. There is still a lot of untapped potential for AR, VR, and MR in environmental applications. The term "automation of forests" entails the ethical infusion of cutting-edge innovation into forests to advance the recent trends utilized for data collection, processing, and surveillance of the forest environment within the scope of research and development. The Internet of Things, Wireless Sensor Network (wsn), Internet of Trees, Deep Learning, as well as other technological advances can be used efficiently to accomplish these goals. A new technology known as the Internet of Forest Things (IoFT) uses distributed smart devices to collect streams of data during monitoring and fire detection to preserve forest sustainability and protect forests from threats. Additionally, deep learning and blockchain technology could enhance smart IoFT device networking, processing, and resource usage.

11.1 Introduction

Nowadays, it is common for participants to exchange information in real time and from any location due to advancements in information gathering and transmission [1]. However, the information is necessary to change the decision-making process in a dynamic way. One industry that is rapidly becoming technologically advanced is the smart forest, which now has sites for data collection, processing, transportation, and analysis. When attempting to make stream processing a trustworthy solution, processing industrial IoT data presents some distinctive obstacles. The IoT industry faces a number of unique challenges regarding data processing, including (1) the large information extraction and handling costs for the huge volume of diverse streaming data produced by a significant number of IoT devices, and (2) the lost and interrupted events caused by data

M. S. Sadiq, Department of Agricultural Economics and Extension, FUD, P.M.B. 7156, Dutse, Nigeria, e-mail: sadiqsanusi30@gmail.com
I. P. Singh, Department of Agricultural Economics, SKRAU, Bikaner, India
M. M. Ahmad, Department of Agricultural Economics and Extension, BUK, Kano, Nigeria
M. Babawachiko, MCA-Google Apps, Suresh Gyan Vihar University, Jaipur, India

https://doi.org/10.1515/9783110785234-011

unpredictability that impede the industry's need for rapid response times [1]. It is more difficult for streaming data processing systems to handle such a considerable increase in streaming rates since these IoT devices send data at a floating rate instead of a constant rate (for instance, irregular events) [2]. Controlling this enormous inflow of data is a very difficult task. The technology involved in stream processing has to be capable of handling enormous amounts of data if the IoT goal is to be fully realized. In response to this need, a few scalable and dependable platforms, such as Apache Spark, Apache Flink, Apache Samza, and Apache Storm, have recently developed, facilitating the real-time improvement of IoT applications [3]. A few complex query languages, including Spark SQL, Flink Table API, KSQL, SamzaSQL, and StromSQL, have been created specifically for performing analytics over streaming data due to the IoT's fast growth. Each of these streaming data queries relies on the idea of windowing to divide constantly unending data streams into discrete data sets with temporal slices of, instance, minutes, seconds, or milliseconds [3]. Forest fire detection is one of the real-world applications for fire data monitoring. The utilization of IoT technologies for remote IoT data gathering to conduct a forest inventory is being increasingly accepted by forest managers [4]. Due to erratic streaming data, dynamic biological changes or climatic variability present a difficulty for IoT-based wildfire monitoring data. In order to avoid unforeseen circumstances having an influence on real-time data, it is imperative to take into account the timeliness of obtained monitoring IoT data.

11.2 Forestry sustainability

While discussing how to utilize energy resources, the term "sustainability" has been increasingly prominent in recent years. The term and meaning of development used most broadly in the Global Environment and Development Commission Report is "sustainability" [5]. Sustainable forest operations (SFO) is a holistic strategy that blends forest activities with goals for socially, environmentally, and economically sustainable development in order to successfully address present and future issues. "A forest is a land area of more than 0.5 ha, with a tree canopy cover of more than 10 %, which is not principally under agricultural or other specified non-forest land use," states FAO [6]. The forest is the hub of the planet's terrestrial biodiversity. In addition to providing livelihoods and reducing carbon emissions and climate change, forests are crucial for sustainably producing food.

The entire area of forests, which will account for 31 % of all land in 2020, is expected to be 4.06 billion ha, while not being regionally or globally distributed. Per person, this translates to 0.52 hectares of forest [5]. The world's forests are divided into four primary regions: tropical (45 %), boreal (27 %), temperate (16 %), and subtropical (11 %). According to Singh et al. [5], Europe accounts for 25 % of the world's forestry area, which is followed by South America (21 %), North and Central America (19 %), Africa (16 %), Asia (15 %), and Oceania (5 %). China, Brazil, the Russian Federation, Canada, and the United

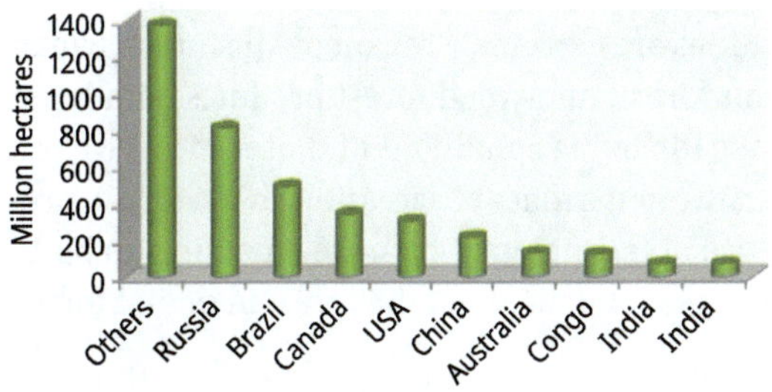

Figure 11.1: Distribution of the major countries with largest forest area coverage in the world.

States of America are the top five countries, with forests covering half of their area (Figure 11.1).

The ecological significance and types of life that can be found in forested environments are referred to as forest biodiversity. In addition to trees, it also comprises a wide variety of forest-specific plants, animals, and microorganisms, as well as their genetic diversity. The degradation of forests and ongoing deforestation have a serious detrimental effect on biodiversity [5]. The permanent removal of trees for logging (or wood extraction for household fuels or coal), agricultural intensification, infrastructure growth, including road construction, and urbanization constitutes deforestation, according to the most basic definition. The climate, precipitation, underground water, air quality, wildlife, and biodiversity are all directly impacted by deforestation. The destruction of forests and the burning of biomass both increase greenhouse gas (GHG) emissions, as per the Intergovernmental Panel on Climate Change [7, 8].

Forest fires are one of the main causes of deforestation since they kill hundreds of trees each year all around the world. The cause of this is the sweltering summers and milder winters. Fires, whether intentionally started or unintentionally, significantly reduce the amount of forest cover. India experienced 520,861 active forest fire occurrences between 2003 and 2017, the majority of which took place in the thick, evergreen, and tropical forests of both the eastern Himalayas and the lower Himalayan states [5]. The primary intentional and unintentional causes of biodiversity loss, declines in the productivity of terrestrial habitats and forest carbon stores, losses in soil quality and resultant crop yields, spikes in air pollutants, and a rising vulnerability to landslides are forest fires. In addition to raising greenhouse gas emissions, soil erosion, floods, species extinction and habitat loss, food insecurity, and biodiversity loss, deforestation has a deleterious impact on the climate [9].

Recent trends in biodiversity and environment are hindering the pursuit of the SDG. Our forests must undergo transformational changes in order to conserve biodiversity and link food production and consumption to the natural world. It is important to distinguish between environmental degradation and the unsustainable production and consumption patterns that accompany economic growth. A study that assessed how dependent locals in nearby forest regions were on fuelwood, fodder, small wood, and bamboo found that about 170,000 villages in India are close to forests [5]. It has also been shown that the people who live in these communities are extremely reliant on the forest, which provides them with food, fuelwood, small forest, non-wood forest products, and bamboo. When formulating policies to improve the living conditions of those who live close to the forest, the lack of information about the items that are accessible in the forest, presented a problem for policymakers. The use of technology will make it possible to build an infrastructure that can supply the whole set of data about the forest in digital form.

11.2.1 Impact of forestry sustainability

Sustainable forestry places a high priority on three outcomes: social sustainability (forests' many uses and non-timber products), ecological sustainability (biodiversity and resilience), and economic sustainability (wood supply) [10]. Finding the best management techniques to forecast how different services, processes, and productions would perform in terms of the environment, economy, and society, is a challenge for the forest operations sector [11]. Sustainable development, which refers to sustaining biodiversity, resilience, productivity, vitality, and significant consequences on economic, environmental, and social activities [12], is the foundation of the idea of sustainable forestry. Forests must be managed sustainably given economic, social, and environmental reasons, as shown in Figure 11.2. Addressing these consequences in terms of

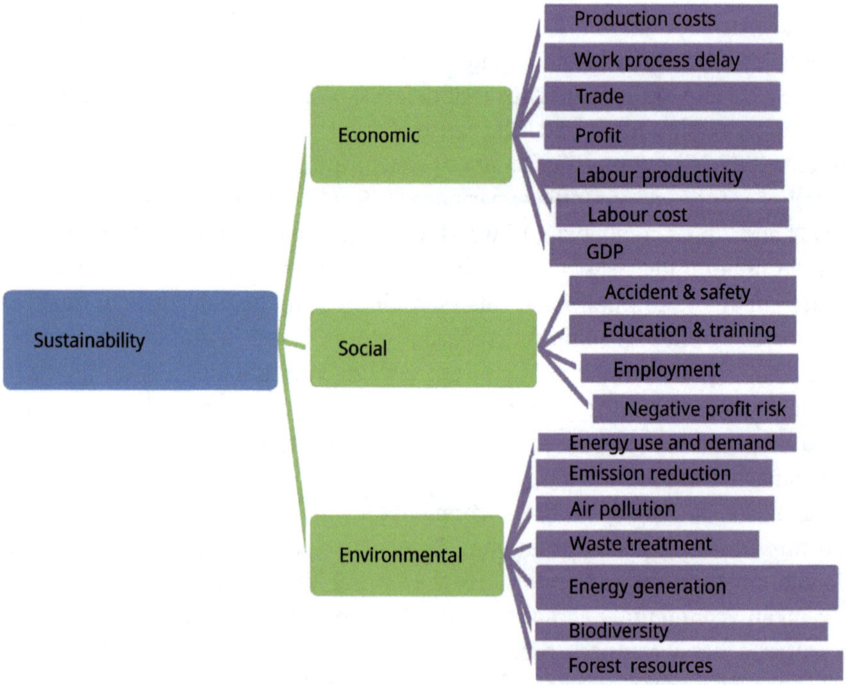

Figure 11.2: Sustainability impacts.

information availability and quality presents a difficulty for developing sustainable forestry [13].

11.3 Internet of things (IoT)

By 2025, there will be 25 billion IoT connections worldwide, according to GSMA intelligence forecasts [5]. Figure 11.3 depicts the growth of IoT devices between 2015 and 2025. IoT stands for the Internet Protocol (IP)-based connection of physical items with a virtual environment [14]. The specific capabilities provided by the IoT components will enhance common applications.

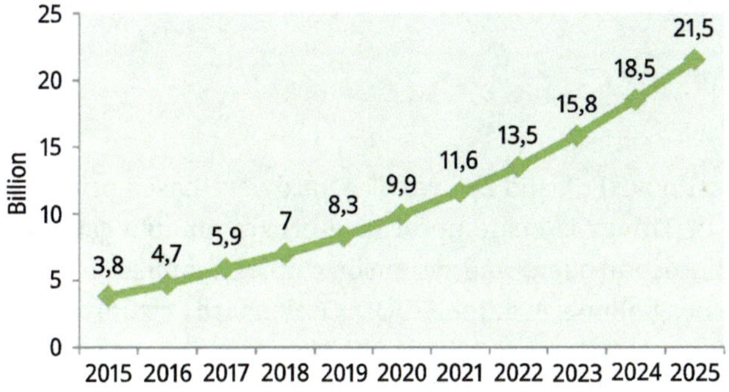

Figure 11.3: Growth of IoT devices between 2015 and 2025.

As depicted in Figure 11.4, the IoT's component parts are described as follows. Identification is required for IoT to categorize and arrange the services according to their needs. Numerous methods, such as pervasive codes and electronic item codes (EPCs), can be used to identify IoT devices [2, 8]. In order to distinguish between an entity's ID and address, it is also necessary to acknowledge IoT entities. The entity's address is its location within the communication network, whereas its ID is its name, such as "H100" for a simple sensor. An object that receives and reacts to input from the physical environment is a sensor. Conventional sensors are capable of measuring environmental factors, processing variables, and digitizing analog signals. However, intelligent sensors are capable of digitizing sensory data.

A number of wireless systems can be employed to establish transmission, which is required for the transfer of sensory data. The Internet of Things makes use of numerous wireless connection methods. Among the communication technologies are IEEE 802.11 g Wireless Fidelity (Wi-Fi), the Global System for Mobile Communication (GSM/GPRS), Zigbee, BLE (Bluetooth Low Energy), 6LoWPAN, RFID, LoRa, Sigfox, NB-IoT, LTE (Long Term

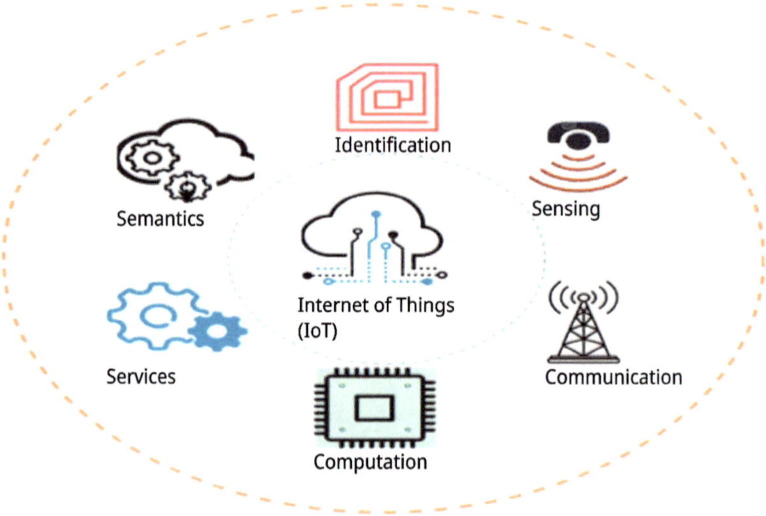

Figure 11.4: Components of the IoT [15].

Evolution), near field communication (NFC), and Z-wave. The hardware-based processing component of the Internet of Things is made up of field-programmable gate arrays (FPGA), microprocessors, microcontrollers, and system on chips. Examples of hardware computing devices include BeagleBone, Arduino, UDOO, Cubieboard FriendlyARM, Raspberry PI, Gadgeteer, Z1, WiSense, Mulle Intel Galileo, & T-Mote Sky. The prominent real-time development frameworks for Internet of Things apps are Tiny OS, Riot OS, and Lite OS. A computer system serves as the Internet of Things' cloud platform. The sensory data is converted to machine-readable form on this platform. These four IoT services are ubiquitous, collaboratively aware, identity-related, and typical data mining services [16].

Services that are pervasive and collaboratively aware employ data templates to make accurate, intelligent decisions in order to provide services to their members at any moment and wherever. Every program contributes real-time items to the virtual environment, demanding their identification, according to identity-related services. Before analyzing and transferring the raw sensory measures to the IoT platform, information aggregation services collect and quantify them. Some of the uses that fall under the four IoT services are smart cities, industrial automation, smart farming, and smart agriculture. In order to deliver services, the semantics function as a route for skillfully obtaining knowledge. By sending requests to the proper resource, semantic reflects the IoT's brain. Platforms like Resource Description Framework (RDF) and Web Ontology Language (OWL) have been advised for the development of semantics [17].

11.4 Applications of Virtual Reality in forestry

Ivan Sutherland's report "The Ultimate Display," which he produced for IFIP in 1965, was the first to propose the idea of VR technology [18–20]. Although the term "virtual reality" (VR) first initially used in the 1980s, the scientific literature has undergone multiple definitional modifications. VR can be characterized as a continuous, three dimensional computer-simulated environment in which users can interact with the 3D produced items and engage themselves in the realistic virtual space, yet there is currently no formal, accepted definition of what VR is. The made-up environment attempts to mirror reality as closely as possible.

The rendering of the virtual environment is accomplished through the use of specialized hardware tools and 3D visualization software. Through visual and acoustic simulations, users can extract information from the 3D models. Virtual Reality (VR), according to Ivan Sutherland, is a window through which a user experiences a virtual environment as though it appeared, felt, and sounded genuine and in which the user might act naturally [21]. Rapid technological advancements in hardware and software systems have improved data preservation, processing, and transmission, as well as the digitization of services as we move into the fourth industrial revolution (I 4.0) [22]. Due in part to this digital revolution, resurgence in VR device development since 2016 has led to more technological improvements.

The forestry industry has recently experimented with VR technology and used the 3D perspective to recreate forested areas. The industry refers to these recreated forests as "Virtual Forests," and they can be used for everything from teaching materials to conducting forestry inventories [21]. When compared to 2D tools such GIS on either a desktop or a laptop environment, Virtual Reality (VR) has the capability to enhance the representation of 3D spatial information, including the topologies of trees inside a forest [21]. In today's increasingly digital environment, the forestry sector must quickly adapt to the new requirements of the Industry 4.0 (I4.0) revolution [22]. The body of research regarding VR applications in the forestry industry is incredibly scant and dispersed, despite ongoing research, growing interest in, and demand for the technology. Some significant topics often covered in that discipline, such as natural ecosystems, forest ecology, forest pathogen, and many others, were absent due to the paucity of literature examining VR in forestry.

Virtual reality for forestry has not yet been properly investigated from a variety of aspects, such as the forestry industry directly, the ecology, tourism, climate change predictions, and ecosystem(s) monitoring, due to its recent expansion and continued development. The proven applications that have previously been created—either through academic research or a few private new tech businesses—are what have been understood and examined thus far. As a result, there is a lot of conjecture about how VR technology may be used, which gives rise to the rhetoric that it is not "completely developed" to meet the needs of large forestry firms and governmental organizations because it is "relatively young."

11.5 Forest digitalization characteristics

Real-time data, real-time surveillance, and real-time forest assessment are made possible by digitalization in the forest. The characteristics that make digitalization possible are as follows:

1. **Sensing technology**: Sensor technologies can create a setting for linking the real and virtual world. In addition to collecting signals, sensors convert them into digital information and process them. Sensors accurately interpret the status of the physical environment at any given time into useful information.
2. **Human-forest interaction**: As humans continue to be an important part of the digitalized forest, it is essential to integrate with the Cyber-Physical Systems (CPS) there. It is important to display data from the forest's CPS on a portable device that helps people make decisions. The development of reliable mobile devices with touch screens should be the main area of focus in this scenario.
3. **Using big data and the cloud**: A significant amount of data is generated during the communication process between the physical system embedded in the forest and must be stored. Cloud computing serves as a platform for digital interaction that stores and displays data. Cloud computing often makes big data analysis easier and more economical in the form of pay-per-use programs. Advanced analytics technology is essential for transforming enormous amounts of data into knowledge that can be used to enhance decision-making.
4. **Artificial intelligence**: The goal of artificial intelligence is to provide technical objects with the capacity to acquire knowledge through experience and observation, the capability to form their own opinions, and the power to take action.

11.6 IoT applications in forestry

Four main categories can be used to categorize Internet of Things (IoT) applications in forestry: monitoring of the forest environment, monitoring of its resources, intelligent management of forest fires, and prevention of illicit logging. For forestry, the application of wireless sensor networks (WSN) in conjunction with the Internet of Things is more effective when it comes to resource and environmental monitoring. Based on the ZIGBEE protocol, Zhang et al. [23] created a collection platform for forest environmental parameters. This platform includes a variety of terminal monitoring devices, including photoresistor (LDR), human infrared sensor (PIR), micro-electromechanical system (MEMS), humidity, water level, gas, and temperature sensors. It specifically aims to increase the network's useful life or improve a sensor's duty cycle. Suciu et al. [24] design for a wireless sensor system that includes a method that efficiently utilizes the resources at hand (but a single node).

Figure 11.5a: LoRa-support solar micro weather station and ICT (Source: [11]).

Figure 11.5b: Node attachement (Source: [11]).

Figure 11.5 illustrates a practical strategy for sustainable forest management developed by Kelvin Hirsch et al. [25] using an ecosystem based on Canadian forest fire prevention. The strategy aims to lessen both the dangers related to using controlled fires and the total area scorched by wildfires by actively and consciously implementing methods for forest management. A system for detecting forest fires was developed by Bolourchi and Uysa [26] and Dener [27] and consists of sensor nodes that are dispersed randomly throughout the forest. Each node has a temperature sensor. Each node has the ability to periodically check for environmental emergencies. If a sensor network recognizes a substantial temperature change by releasing a memory buffer with its values obtained, the information will be displayed on the PC website page as well as the mobile telephony page. A technique to stop the smuggling of valuable trees like red and sandalwood in forested areas was provided by Suguvanam et al. [28]. Their model consists of three components: one-tree unit, a sub-server, and a ranger in the forest.

11.7 Internet of Forest Things (IoFT)

The Internet of Forest Things (IoFT) refers to the use of smart devices positioned across forests for governance, monitoring, and the discovery and protection of fires, representing, a new generation of the Internet of Things. Figure 11.6 shows a broad overview of the forest fire scenario, where the deployed IoFT is used to monitor meteorological variables like temperature, humidity, CO (carbon monoxide), and CO_2 (carbon dioxide). Apache Kafka collects and makes use of the sensory meteorological data as a robust message queuing system. The acquired data (i. e., potential fire incidents) is then provided to the search engine running on top of the flow processing system. The risk of a fire igniting varies greatly throughout the day, which causes a delay in reporting fires to the forest department's authority. Moreover, because of the delay, the forest department will have the power to dispatch drones and firemen to a burning forest to sound the sirens later.

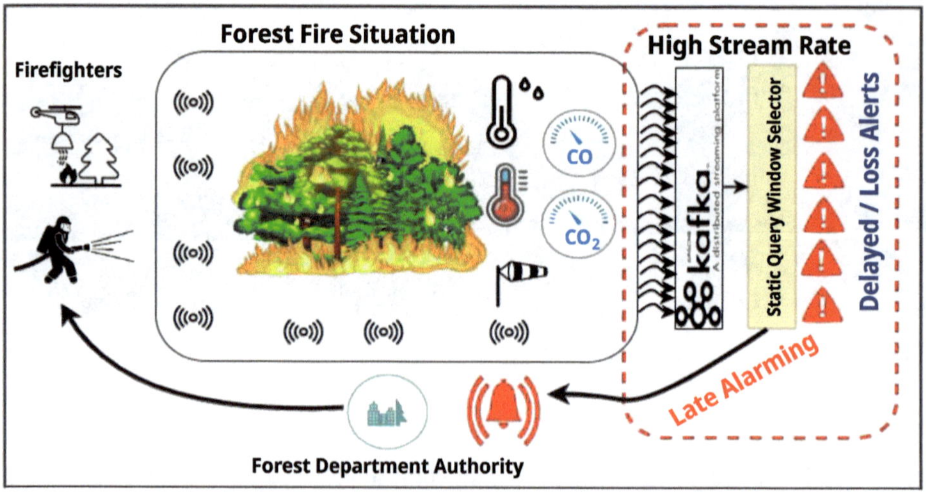

Figure 11.6: IoFT using weather monitoring data, fire detection and protection (Source: [3]).

11.8 The Internet of Trees

The "Internet of Trees" concept pertains to the incorporation of internet-based components into trees for the purpose of online factor detection, communication, and monitoring in forests. The usage of internet-based components for real-time environmental indicator monitoring, fire accident identification, and forest degradation detection is referred to as the "Internet of Trees."

We heard a lot about the Internet of Things (IoT), which uses IP to track, monitor, and connect with objects [29]. Prior research has not paid much attention to monitoring environmental variables including temperature, wind speed, relative humidity, as well as carbon dioxide. The existing studies in the field have largely focused on identifying and keeping track of forest fires (CO_2). The Amazon forest caught fire in 2019 as a consequence of careless and deliberate human conduct. The Amazon rainforest plays the essential role as the planet's lungs and produces a significant quantity of oxygen by measuring CO_2. The Amazon forest fires increased carbon monoxide (CO) emissions, and the CO_2 level in the air is dangerous for human health, according to study from the Copernicus Climate Change Service (C3S) [30].

11.8.1 Monitoring of environmental parameters

Various trees have sensor nodes built into them, enabling the real-time perception of the forest's environmental factors. The sensor node frequently incorporates a wireless network module and environmental sensors. The architecture includes a wireless personal network (WPAN) to aid with wireless connectivity, which is the major challenge in a forest environment. A wireless communication mechanism in the sensor node for transmitting environmental information from the forest environment is believed to be an IEEE 802.15.4 WPAN-based Zigbee module. The gateway node may use IP to log through into a cloud server because it has internet access and gets data *via* Zigbee. Figure 11.7 depicts the overall structure.

The ability to collect and transmit sensory information from the forest toward a cloud server is made possible by sensor technology and the wireless connection protocol (Kim et al., 2020 [49]). According to Singh et al. [5] and Al-Fuqaha et al. [15], raspberry pi 3 based forest surveillance is recommended for monitoring and sharing environmental parameters including temperature, humidity, and gas concentrations utilizing IP. The following gases in the forest are tracked by an IP and Transmission Control Protocol (TCP)-based system: carbon dioxide (MQ 135), hydrogen (MQ 2), methane (MQ 4), and carbon monoxide (MQ 9) [16]. The environmental effects of the forests are monitored and communicated using just a LoRa-based sensor network plus gateway (Figure 11.8). Low Earth Orbit satellites and IoT are combining to monitor and communicate with the Indonesian forest [32]. There are drawbacks to the Zigbee design for monitoring forests, such as limited transmission range and a dearth of analytics at the gateway node [33]. Thanks to the advancement of the LoRa communication module, data may now be reliably and securely delivered over a long distance.

Sensory data is transmitted using a LoRa communication module, which uses less power. Low-interference data from sensors is transmitted across the forest using LoRa

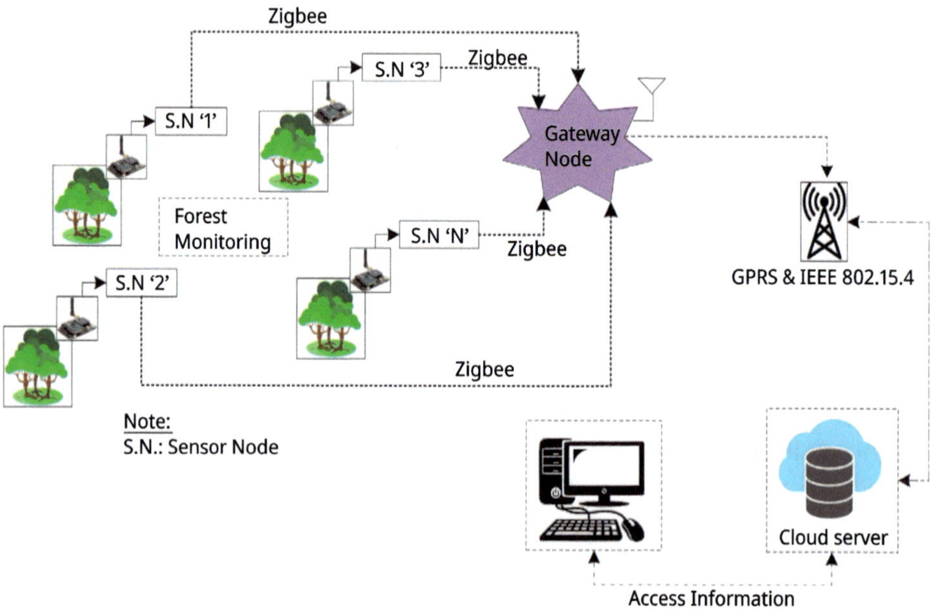

Figure 11.7: Forest monitoring architecture [31].

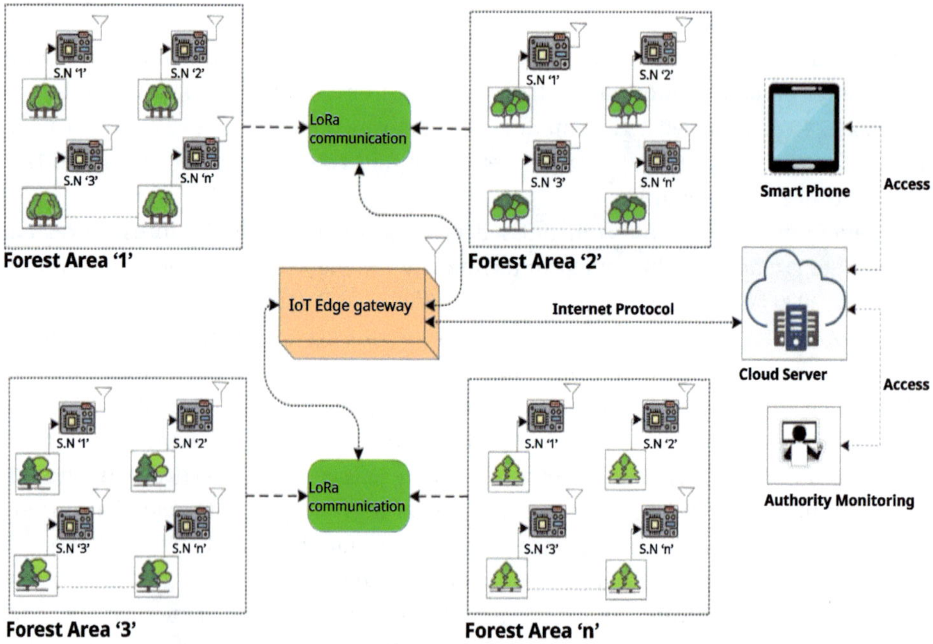

Figure 11.8: Real-time forest ecosystem monitoring with LoRa and edge gateways (Source: [5]).

technology. The edge gateway node receives the data from the LoRa-based sensor node. IoT frequently combines cloud servers for storage and analytics on sensory data to provide comprehensible metadata. However, in the situation of the forest viewpoint, a quick decision must be taken to prevent a large-scale forest fire. In this situation, edge computing technology satisfies the demand for rapid decision-making at the edge devices. An analytics function can be performed at the gateway node thanks to the IoT edge gateway's use of edge computing. After giving the analytics, this gateway node transfers the accurate data through IP to the cloud server.

11.8.2 Monitoring and tracking of fires

The Himalayan mountainous region is extremely sensitive to wildfire mishaps because of the high-density forest [34]. According to a Forest Study of India (FSI) fact sheet [32], 54.40 %, 0.7.49 %, and 2.41 % of India's forest cover is exposed to frequent fires, very frequently to fires, and at elevated levels, respectively. According to the FSI, the primary fire-prone areas in India are in the north-east and the center, and a fifth of the forest is at risk of burning. Almost 90 % of forest fires in India are caused by humans, emphasizing the importance of implementing appropriate mitigation measures and establishing sensitive forest fire zones to prevent further escalation of forest fire concerns in the natural environment.. Remote sensing and geoinformation system (GIS) have been used to spatially map the fire-prone areas to detect wildfire incidents [35]. An effective method for identifying the real-time components of a forest environment is necessary, because the use of remote sensing and GIS technology for forest monitoring only offers geographical data. A 433 MHz limited communication standard and IEEE 802.15.4 are used to put up a low-power sensor for ecological monitoring. Fires are detected and monitored via a Wi-Fi equipped sensor network [36]. For the purpose of spotting wildfires and keeping track of environmental variables in woods, a cloud/fog computing solution based on Zigbee and IEEE 802.15.4 connectivity is proposed [37, 38]. A novel system for detecting fires and notifying a distant server about them is integrated with an STM32 controller, a Zigbee module, plus GPRS (Global Packet for Radio Service) [39]. For the purpose of detecting wildfires by changing the crown and surface fire, a wireless acoustical monitoring framework Internet of Things is developed [23]. Several methods and technologies have been proposed for real-time monitoring and tracking of forest fires, including low-power sensors, Wi-Fi-equipped sensor networks, cloud/fog computing solutions, and IoT-based wireless acoustical monitoring frameworks.

Up until now, forests have been monitored via the Internet of Things, but illegal logging is another factor accelerating forest deterioration. A surveillance system based on the Internet of Things is recommended for stopping smuggling, detecting forest fires, and communicating with forest authorities through wireless communication protocols [40]. The use of internet-based modules improves the ability of any site with an internet connection to monitor forest events in real-time.

11.9 Internet of Minor Forest Produce (IoMfP)

Minor forest products (MFP), also known as non-timber forest produce (NTFP), are an essential source of income for many people living in and around forests. MFPs provide essential food, nutrition, medical supplies, and financial support to several indigenous groups. The forestry sector has significant potential to improve the living standards of forest-dependent communities, especially indigenous tribes, through sustainable harvesting, production, value addition, and marketing of MFPs. The products generated by minor forest producers are depicted in Figure 11.9. According to the Minor Forest Produce (MFP) 2020 report, 100 million people live in forests and depend on the products, medicines, and cash they produce. All non-timber forest products fall under the category of Minor Forest Produce, which encompasses all non-timber forest products, including fodder for livestock, and vegetables.

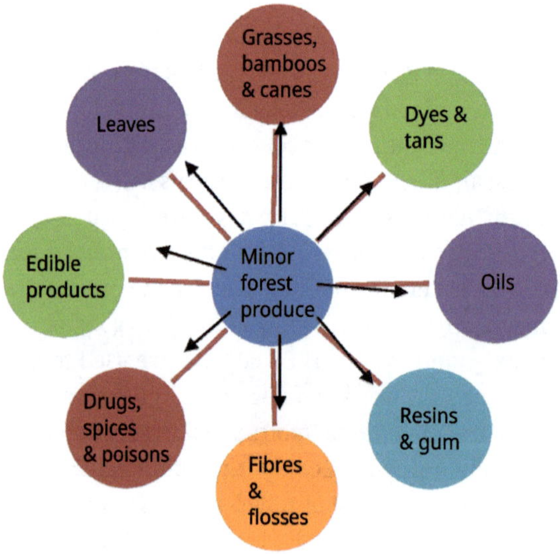

Figure 11.9: Minor forest products.

11.9.1 Grass

Grass is used in various applications, including the production of paper, matting, cordage, and cooling screens. A substantial portion of grassland is utilized as feed and thatch. The most significant grass that provides the necessary raw materials for the paper mills is *Eulaliopsis binata* (Sabai grass) is a significant grass species providing essential raw materials for paper mills. *Saccharum munja* (Munj grass) is used for making benches and stools, while *Chrysopogon zizanioides* (vetiver or khus grass) is used for creating cooling panels.

11.9.2 Bamboo

Bamboo (*Bambusa tulda*) is an evergreen perennial flowering plant from the *Poaceae* grass family's Bambusoideae subfamily. It is tall, woody, and perennial. Around 100 bamboo species grow in Indian forests, covering over one lakh square kilometers (100.000 km^2). Bamboo provides low-cost materials for roofing, basketry, walls, flooring, and mattresses for tribal people. Cane grass, or members of the genus *Arundinaria*, is a tall perennial grass with flexible, woody stalks, primarily used for ropes, belts, furniture, and sporting goods. It grows mainly in the wet forests of India.

11.9.3 Dyes and Tans

The tanoak or tanbark-oak (*Lithocarpus densiflorus*) plant tissue is mostly used in leather production. Dyes extracted from various plants, such as *Lawsonia inermis, Indigofera tinctoria,* and *Isatis tinctoria,* have their dyes extracted, which are then used to color textiles, fruit, pharmaceuticals, and cosmetics. Many plants and trees in Indian forests produce oils that are used in soap, cosmetics, and pharmaceutical preparations.

11.9.4 Resins and gum

Gum (*Eucalyptus camaldulensis*) seeps from tree stems either spontaneously or partially, due to bark, wood, or burning exposure. Examples of gum trees found in forests include Babul (*Acacia nilotica*), Khair (catechu), Kullu (*Sterculia urens*), Dhawra (*Anogeissus latifolia*), Palas (*Butea monosperma*), Semal (*Bauhinia retusa*), Lendia (*Lannea coromandelica*), and Neem (*Azadirachta indica*). Chir pine (*Pinus roxburghii*) is mainly used for extracting resin, which is used in various industries like, waterproofing, acrylic, paint, varnish, rubber, paper, linoleum, sticky tape, oils, and greases.

11.9.5 Fibres

Natural fibers, also known as lignocellulosic fibers, are plant and fiber extracts used for making ropes due to their strength. Fishing nets are made from fibers of akund floss (AK), which is also used for stuffing pillows and mattresses.

11.9.6 Leaves

Various tree leaves serve different purposes, such as tendu leaves for wrapping bidis and *Bauhinia vahlii* leaves for making plates and leaf cups used by sweet vendors as packaging. Many plant and tree parts, including stems, seeds, leaves, and branches, are used to produce medicines. Numerous herbs and spices such as *Eugenia caryo* (turmeric), *Piper nigrum L.* (black pepper), *Elettaria cardamom* (cardamom), *Curcuma longa phyllata* (clove), *Cinnamomum zeylanicum* (cinnamon), and *Zingiber officinale* (ginger), are used in cooking to enhance flavors. Some forest-produced poisons have medicinal applications; they can treat ailments if eaten in small amounts, namely: *Aconitum napellus* (Aconite), *Datura stramonium* (datura), *Strychnos nux-vomica* (strychnine), *Cannabis sativa, Cannabis indica* (ganja), and others. Edible products include flowers, fruits, and leaves from various trees and plants.

The Internet of Things (IoT) is currently one of the key technologies capable of digitizing information in real-time. By leveraging, digitalization and IoT the global exploration of minor forest products can be enhanced, thereby boosting marketing and sales. The Government of India (GoI), has set a goal to improve the supply chain infrastructure and increase the commercialization of minor forest products, while simultaneously sustaining the livelihood of forest dwellers. IoT plays a crucial role in supporting and aiding these objectives.

11.9.7 MFP cycle integration with IoT

The MFP cycle facilitates the transfer of forest-based raw materials to the finished product. Components that make up the MFP cycle include gathering, processing, quality enhancement, packaging, marketing, retail and consumers. Natives living near the forest often collect the raw materials. To supply the raw materials and obtain the income necessary to support their families, tribal people go to the local processing facility. The acquired raw material is transformed into usable products during the processing stage. Value addition involves adding features to a product to offer customers a superior and distinctive product. The packaging stage is crucial to preserve the product for an extended period. The package labeling includes information about the ingredients used in the product. Marketing involves promoting and selling products using various strategies. Retailers distribute products to customers in small quantities for personal use rather than resale. The process is depicted in Figure 11.10 and the entire cycle is carried out on paper documents, allowing for manipulation of data from collection to consumers.

Due to the lack of a tracking system during the MFP cycle, mediators can easily falsify data regarding the products. Modern consumers are curious to learn a product's history, from raw materials to the finished product. Improving the current MFP cycle is necessary, which requires implementing IoT-based hardware. The deployment of IoT-based sensor technology covers the MFP cycle, encompassing all activities occurring within forests from the generation of basic resources to the sale of marketable forest-based products [41]. These IoT enabled solutions, improve the MFP cycle's efficiency to address the issue of interdependencies across multiple organizations.

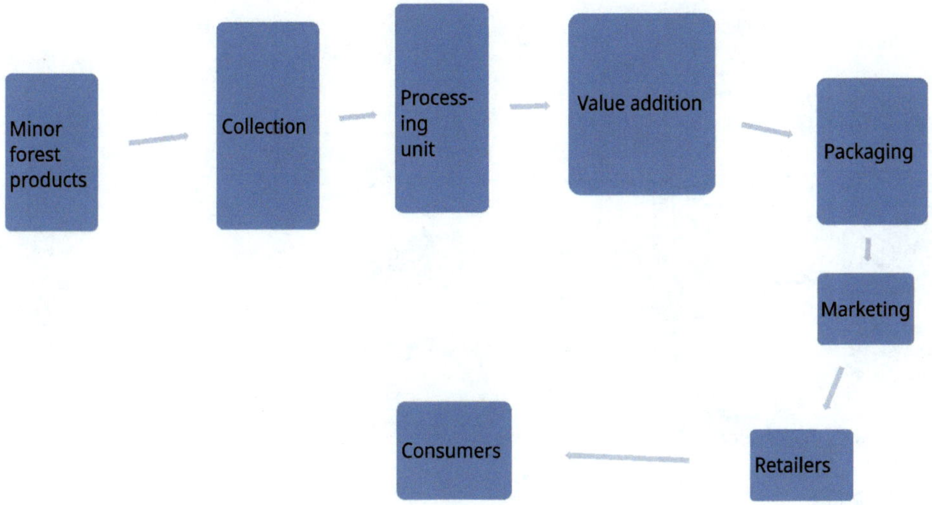

Figure 11.10: MFP without IoT.

Higher authorities can now remotely monitor activities in real-time via IP due to the deployment and incorporation of IoT-based devices at the processing unit. In this system, each product is assigned a unique identity that enables tracking of all relevant product data, including the product's origin, the types of raw materials used, its validity, and all certificates. Barcodes and RFID are excellent identification technologies for real-time tracking due to their widespread use and lightweight design [33]. Figure 11.11 shows a multifunction printer (MFP) with an embedded processing unit and packaging unit that use various wireless technologies and sensors to convert into digital monitoring as an example of an Internet of Things (IoT) paradigm. The sensor node in this scenario, incorporated into the processing and packaging facilities, has a camera, an RFID reader, a barcode reader, a LoRa modem, and GPS attached to it. Environmental sensors monitor factors like temperature, humidity, and smoke within the packaging/processing unit. These variables are detected by environmental sensors, which then communicate with a nearby IoT-based gateway. The product is labeled with a "barcode" and a "RFID" tag within the packaging unit. If any indicators exceed the threshold limit, the gateway immediately notifies the cloud server. With the sensor module connected to the RFID and barcode scanners, all data scanned by these devices is transmitted to the cloud server through LoRa and internet connectivity. All product-related activities are digitally recorded on a cloud server. Authorities, manufacturers, retailers and consumers can monitor and trace products continuously and in real-time thanks to the cloud server's data accessibility. With internet access, authorities can assess a product's quality remotely and instruct staff at a processing or assembly facility on maintaining the product's environmental standards. The development of digital platforms and the availability of product data in digital format have enabled global advertising and mar-

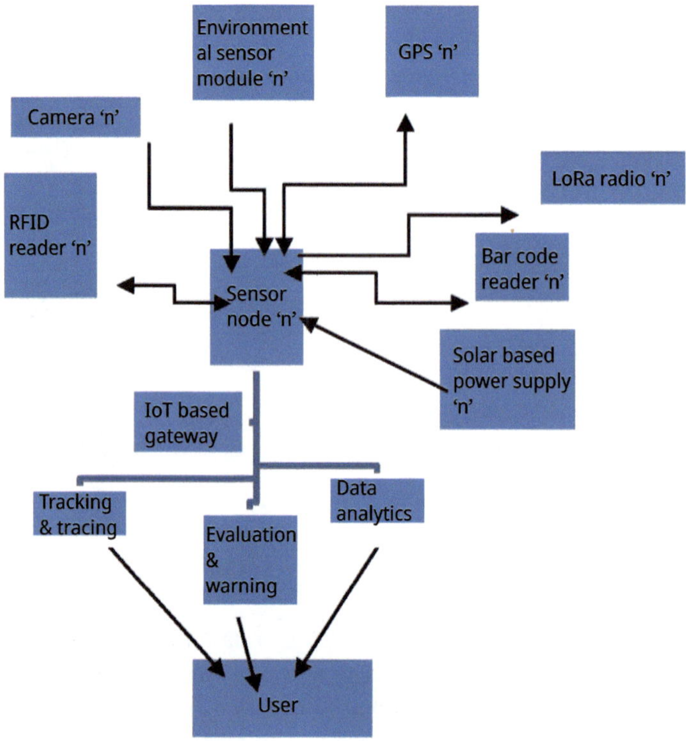

Figure 11.11: MFP with IoT.

keting of a product. The use of cloud servers facilitates the creation of an accountability platform for upgrading material flows at various hubs.

11.10 Internet of Wild Things (IoWT)

Wildlife refers to animals, plants, and other living organisms that inhabit their natural environments. Species and their habitats play a significant role in ecological and biological processes which are essential to life itself. Numerous connections between plants, animals, and microbes are necessary for the biosphere to function, protect, and enhance human life. Wild animals participate in various ecological processes, such as pollination, germination, seed dispersal, soil structure formation, nitrogen cycling, scavenging, ecosystem preservation, waste management, and pest control. Deforestation, poaching, and hunting, however, are the primary causes of wildlife endangerment as they damage habitats. Consequences of endangered species include an imbalanced ecology, loss of biodiversity, and a disrupted food chain.

11.10.1 Incorporating wildlife modules from the internet

IoT offers potential solutions to address the issue of wildlife endangerment. The ability to create wireless sensor networks, often in challenging and remote environments, provides new tools for monitoring endangered animals and their habitats and ensuring their protection [42]. For example, IoT-based collars that detect pulse rate and location have been installed on rhinos to help conserve this vulnerable species. South Africa is home to 70 % of the remaining wild rhino populations. The use of IoT-based collars allows real-time animal health monitoring, which in turn facilitates quick emergency response by wildlife authorities. IoT in Animal Healthcare (IoTAH) monitors animal health using biosensors and communication protocols [43, 44]. This technology provides an actual health status and a disease prognosis. It is recommended to use a collar with LoRa and BLE capabilities to track and map the locations of wild animals [45]. A vision-based IoT device called "Where is the Bear" is used in the UCSB Sedgwick Reserve to detect wild animals using image processing techniques.

11.10.2 IoT-based real-time wildlife monitoring equipment

The Zoological Society of London (ZSL) has developed a monitoring system called Instant Detect, which enables remote tracking of animal behavior and environmental changes, as well as early detection of illegal poaching activities. The Instant Detect 2.0 system (Figure 11.12) combines sensor nodes, cameras, and low-power wide-area networks to sense, record, and transmit real-time images to the base station. An embedded camera in the anti-poaching cutting-edge device TrailGuard AI (Figure 11.13 and 11.14) monitors poaching activity and sends via satellite modems and the L-band network., IoT-based

Figure 11.12: Instant detect 2.0 [46, 5].

Figure 11.13: TrailGuard AI [47, 5].

Figure 11.14: Conservation technology [46].

collars that detect pulse rate and location have been installed on rhinos to help conserve this vulnerable species, as. South Africa is home to 70 % of the remaining wild rhino populations. Based on this chip, the Sxtreo T51 PDA is a portable electronic device based on this chip, which the Odisha government in India uses to protect forests.

11.11 Difficulties with forest digitalization

Implementing cutting-edge digitalization in the forestry industry involves integrating all information related to forests on local, national, and international scales through a sophisticated digital network [48]. The following factors make the implementation of digitization difficult in the forest sector.

11.11.1 Connectivity

Connectivity is crucial for the implementation of a digital network [49]. Communication between sensor nodes and servers is facilitated by connectivity. In a forest setting, hills and uneven terrain pose challenges to connectivity. Previous research transmitted forest sensor data using IEEE 802.15.4 Zigbee and IEEE 802.11 g Wi-Fi modules with a maximum transmission range of 50–100 meters [50]. Implementing Wi-Fi modules in forests is difficult since they require constant internet connectivity.

11.11.2 Real-time sensing

Real-time sensing enables the detection of events such as illegal tree logging, catastrophic events, and changes in forest cover. Only a few organizations currently use real-time sensor technologies to protect species from extinction and poaching. Sensing technologies should be more widely applied in to gather environmental data about forests and detect unexpected events in vegetation and forest cover.

11.11.3 Economically viable infrastructure

Forest range and authorities need a cost-effective infrastructure, such as handheld devices with a cloud-based GUI, to monitor .forest operations in real-time without constantly visiting the forest. A handheld device is portable, easy transportable tool that assists forest authorities in their routine surveillance.

11.11.4 Innovation in IoT device development

Advanced sensor technology and wireless communication protocols enable innovative new tools for monitoring wildlife and forests. However, only private organizations monitoring wildlife on their property can take advantage of IoT devices' innovation. A limited number of commercial companies have developed various tools for real-time animal tracking and health monitoring. Edge-based IoT devices have also been developed for identifying poachers. Therefore, further innovation is needed to create affordable and reliable devices for forests and wildlife.

11.11.5 Collecting energy from sensor nodes

Sensor nodes typically run on batteries to sense and transmit data. It is a time-consuming and inconvenient task, as officials must travel to the nodes to check their battery life. Solar energy harvesting could be a potential solution to the sensor nodes' energy shortage. The sensor node could use solar energy collecting as a backup energy source to prevent data loss and connection interruptions.

11.11.6 Improving the quality of life for tribal populations

Despite living in areas rich in natural resources, many tribal communities in India face significant discrimination and inequity. The exploitation of natural resources, such as trees due to the shift in Indian culture and industries, has forced local populations to move to the periphery. A skill-development program focused on forest conservation and training in the deployment of IoT-based devices in forests could improve the livelihoods of tribal communities. This would include the successful implementation of an IoT-based forest monitoring system, and the improvement of tribal population's technological literacy.

11.11.7 Innovative MFP marketing

Stakeholders' primary marketing strategies include product demonstrations, newspaper advertisements, trade shows, and the internet. To ensure the effectiveness of these strategies, stakeholders must conduct formal market research to understand consumer needs and preferences. Digital surveys and feedback are utilized by stakeholders to track customer needs.

11.11.8 Computer vision node

In contrast to sensors, which rely on the physical environment and transmit data in the form of electrical packets, the computer vision node can function as a cross-check system. For instance, in predicting forest fires, the vision node would use its internal machine learning algorithm, such as a support vector machine, to provide an accurate prediction of the water content in nearby plants. By comparing this to the information received from sensor nodes, like wind speed and temperature fluctuations, it can be used to predict the likelihood and direction in which a fire might spread. Similarly, for flora analysis, the vision node could employ photometric stereo to monitor and predict plant growth in relevant environments.

11.11.9 Inclusion of artificial intelligence and deep learning

The data capture technologies employed by the cloud server provide a vast amount of forest data. Deep learning (DL) and machine learning (ML) techniques can be applied to extract insights from this data. By analyzing images recorded and saved on the server, DL and ML can predict the appearance of anomalies in the forest environment.. Potential applications of integrating ML and DL in forestry include assessing variability in flora, predicting and detecting fire, and identifying diseases.

11.12 Conclusion

Forests are a crucial natural resource that helps maintain the balance of the environment. Forest degradation leads to increased greenhouse gas emissions, negatively impacting earth's living creatures. The primary causes of forest destruction are fire incidents and tree harvesting. Wildlife is also closely associated with forests, and due to forest degradation and hunting-related poaching, it now faces the risk of extinction. When wildlife is endangered; the ecological system becomes unbalanced. Moreover, people living in forest areas struggle to make a living as a result of the forest's deterioration. Therefore, there is a need for real-time technology that can detect occurrences in the forest environment to improve vegetation, protect endangered species, and enhance the quality of life for local communities. Real-time monitoring can be implemented using IoT by integrating IoT devices into the forest. The IoT approach for the MFP cycle allows authorities to assess the entire cycle on a digital platform which can be used to develop policies aimed at boosting the sales of MFP products.

Bibliography

[1] S. K. Lakshmanaprabu, K. Shankar, M. Ilayaraja, A. W. Nasir, V. Vijayakumar, and N. Chilamkurti, "Random Forest for Big Data Classification in the Internet of Things Using Optimal Features," International Journal of Machine Learning and Cybernetics, vol. 10, no. 10, pp. 2609–2618, 2019.

[2] D. Liu, H. Zhen, D. Kong, X. Chen, L. Zhang, M. Yuan, and H. Wang, "Sensors Anomaly Detection of Industrial Internet of Things Based on Isolated Forest Algorithm and Data Compression," Scientific Programming, vol. 2021, 2021.

[3] R. Sahal, S. H. Alsamhi, J. G. Breslin, and M. I. Ali, "Industry 4.0 Towards Forestry 4.0: Fire Detection Use Case," Sensors, vol. 21, no. 3, p. 694, 2021.

[4] Y. Qi and X. Gao, "Equivalence Assessment Method of Forest Tourism Safety Based on Internet of Things Application," Computational Intelligence and Neuroscience, vol. 2022, 2022.

[5] R. Singh, A. Gehlot, S. V. Akram, A. K. Thakur, D. Buddhi, and P. K. Das, "Forest 4.0: Digitalization of Forest Using the Internet of Things (IoT)," Journal of King Saud University: Computer and Information Sciences, 2021.

[6] K. Mehta, S. Sharma, and D. Mishra, "Internet-of-Things Enabled Forest Fire Detection System," In 2021 Fifth International Conference on I-SMAC (IoT in Social, Mobile, Analytics and Cloud) (I-SMAC), pp. 20–23, IEEE, November 2021.

[7] E. Marchi, W. Chung, R. Visser, D. Abbas, T. Nordfjell, P. S. Mederski, et al., "Sustainable Forest Operations (SFO): A New Paradigm in a Changing World and Climate," Science of the Total Environment, vol. 634, pp. 1385–1397, 2018.

[8] L. Liu, J. Wang, F. Wang, and X. Yang, "The Impact of the Planting of Forest Biomass Energy Plants Under the Embedded Internet of Things Technology on the Biodiversity of the Local Environmental Ecology," Environmental Technology and Innovation, vol. 24, 101894, 2021.

[9] A. Sungheetha and R. Sharma, "Real Time Monitoring and Fire Detection Using Internet of Things and Cloud Based Drones," Journal of Soft Computing Paradigm (JSCP), vol. 2, no. 03, pp. 168–174, 2020.

[10] J. P. Siry, F. W. Cubbage, K. M. Potter, and K. McGinley, "Current Perspectives on Sustainable Forest Management: North America," Current Forestry Reports, vol. 4, no. 3, pp. 138–149, 2018.

[11] S. T. Chen, C. C. Hua, and C. C. Chuang, "Application of IOT to Forest Management Taking Fushan Botanical Garden as an Example," Planning and Management, vol. 63, no. 3, pp. 481–499, 2021.
[12] S. T. Chen, C. C. Hua, and C. C. Chuang, "Forest Management Using Internet of Things in the Fushan Botanical Garden in Taiwan," Journal of Advance Artificial Life Robot, vol. 2, p. 2795, 2021.
[13] D. Tuomasjukka, S. Martire, M. Lindner, D. Athanassiadis, M. Kühmaier, J. Tumajer, et al., "Sustainability Impacts of Increased Forest Biomass Feedstock Supply – A Comparative Assessment of Technological Solutions," International Journal of Forest Engineering, vol. 29, no. 2, pp. 99–116, 2018.
[14] I. M. Wildani and I. N. Yulita, "Classifying Botnet Attack on Internet of Things Device Using Random Forest," In IOP Conference Series: Earth and Environmental Science, vol. 248(1). page 012002, IOP Publishing, March 2019.
[15] A. Al-Fuqaha, M. Guizani, M. Mohammadi, M. Aledhari, and M. Ayyash, "Internet of Things: A Survey on Enabling Technologies, Protocols, and Applications," IEEE Communications Surveys and Tutorials, vol. 17, no. 4, pp. 2347–2376, 2015.
[16] X. Xiaojiang, W. Jianli, and L. Mingdong, "Services and Key Technologies of the Internet of Things," ZTE Communications, vol. 8, no. 2, pp. 26–29, 2020.
[17] B. Yan, J. Yu, M. Yang, H. Jiang, Z. Wan, and L. Ni, "A Novel Distributed Social Internet of Things Service Recommendation Scheme Based on LSH Forest," Personal and Ubiquitous Computing, vol. 25, no. 6, pp. 1013–1026, 2021.
[18] M. P. White, N. L. Yeo, P. Vassiljev, R. Lundstedt, M. Wallergård, M. Albin, and M. Lõhmus, "A Prescription for "Nature" – The Potential of Using Virtual Nature in Therapeutics," Neuropsychiatric Disease and Treatment, 2018.
[19] J. Huang, M. S. Lucash, R. M. Scheller, and A. Klippel, "Visualizing Ecological Data in Virtual Reality," In 2019 IEEE Conference on Virtual Reality and 3D User Interfaces (VR), pp. 1311–1312, IEEE, 2019.
[20] J. Huang, M. S. Lucash, M. B. Simpson, C. Helgeson, and A. Klippel, "Visualizing Natural Environments From Data in Virtual Reality: Combining Realism and Uncertainty," In IEEE Conference on Virtual Reality and 3D User Interfaces (VR), pp. 1485–1488, IEEE, 2019.
[21] M. Perez-Huet, Global Literature Review on the Applications of Virtual Reality in Forestry. Being a *Master's Thesis* submitted to the Faculty of Science and Forestry, University of Eastern Finland, Finland, p. 9.
[22] F. Müller, D. Jaeger, and M. Hanewinkel, "Digitization in Wood Supply – A Review on How Industry 4.0 Will Change the Forest Value Chain," Computers and Electronics in Agriculture, vol. 162, pp. 206–218, 2019.
[23] S. Zhang, D. Gao, H. Lin, and Q. Sun, "Wildfire Detection Using Sound Spectrum Analysis Based on the Internet of Things," Sensors, vol. 19, no. 23, p. 5093, 2019.
[24] G. Suciu, R. Ciuciuc, A. Pasat, and A. Scheianu, "Remote Sensing for Forest Environment Preservation," In World Conference on Information Systems and Technologies, pp. 211–220, Springer, Cham, April 2017.
[25] K. Hirsch, V. Kafka, C. Tymstra, R. McAlpine, B. Hawkes, H. Stegehuis, et al., "Fire-Smart Forest Management: a Pragmatic Approach to Sustainable Forest Management in Fire-Dominated Ecosystems," Forestry Chronicle, vol. 77, no. (2, pp. 357–363, 2001.
[26] P. Bolourchi and S. Uysa, "Forest Fire Detection With Wireless Sensor Networks," In Fifth International Conference on Computational Intelligence, Communication Systems and Networks, 2013.
[27] M. Dener, Y. Özkök, and C. Bostancioglu, "Fire Detection Systems in Wireless Sensor Networks," In World Conference on Technology, Innovation and Entrepreneurship, Social and Behavioral Sciences, vol. 195, pp. 1846–1850, 2015.
[28] K. R. Suguvanam, R. Senthil Kumar, S. Partha Sarathy, K. Karthick, and S. Raj Kumar, "Innovative Protection of Valuable Trees From Smuggling Using Rfid and Sensors," International Journal of Innovative Research in Science, Engineering, and Technology, vol. 6, no. 3, pp. 3836–3845, 2017.
[29] R. Tomar and R. Tiwari, "Information Delivery System for Early Forest Fire Detection Using Internet of Things," In International Conference on Advances in Computing and Data Sciences, pp. 477–486, Springer, Singapore, 2019.

[30] T. O. Assis, A. P. D. De Aguiar, C. Von Randow, D. M. de Paula Gomes, J. N. Kury, J. P. H. Ometto, and C. A. Nobre, "CO_2 Emissions From Forest Degradation in Brazilian Amazon," Environmental Research Letters, vol. 15, no. 10, 104035, 2020.
[31] F. Wu, X. Lv, and H. Zhang, "Design and Development of Forest Fire Monitoring Terminal," In Proceeding of International Conference on Sensors Networks Signal Process, pp. 40–44, 2018.
[32] J. Ratnam, S. K. Chengappa, S. J. Machado, N. Nataraj, A. M. Osuri, and M. Sankaran, "Functional Traits of Trees From Dry Deciduous "Forests" of Southern India Suggest Seasonal Drought and Fire Are Important Drivers," Frontiers in Ecology and Evolution, vol. 7, p. 8, 2019.
[33] C. K. Wu, K. F. Tsang, Y. Liu, H. Zhu, Y. Wei, H. Wang, and T. T. Yu, "Supply Chain of Things: A Connected Solution to Enhance Supply Chain Productivity," IEEE Communications Magazine, vol. 57, no. 8, pp. 78–83, 2019.
[34] S. Sannigrahi, F. Pilla, B. Basu, A. S. Basu, K. Sarkar, S.. Chakraborti et al., "Examining the Effects of Forest Fire on Terrestrial Carbon Emission and Ecosystem Production in India Using Remote Sensing Approaches," Science of the Total Environment, vol. 725, 138331, 2020.
[35] H. Abedi Gheshlaghi, B. Feizizadeh, and T. Blaschke, "GIS-Based Forest Fire Risk Mapping Using the Analytical Network Process and Fuzzy Logic," Journal of Environmental, 2020.
[36] S. Pareek, S. Shrivastava, S. Jhala, J. A. Siddiqui, and S. Patidar, "IoT and Image Processing Based Forest Monitoring and Counteracting System," In 4th International Conference on Trends in Electronics and Informatics (ICOEI)(48184), pp. 1024–1027, IEEE, 2020.
[37] A. Tsipis, A. Papamichail, I. Angelis, G. Koufoudakis, G. Tsoumanis, and K. Oikonomou, "An Alertness-Adjustable Cloud/Fog IoT Solution for Timely Environmental Monitoring Based on Wildfire Risk Forecasting," Energies, vol. 13, no. 14, p. 3693, 2020.
[38] S. Srividhya and S. Sankaranarayanan, "IoT–Fog Enabled Framework for Forest Fire Management System," In Fourth World Conference on Smart Trends in Systems, Security and Sustainability (WorldS4), pp. 273–276, IEEE, July 2020.
[39] X. Yunjie, "Wireless Sensor Monitoring System of Canadian Poplar Forests Based on Internet of Things," Artificial Life and Robotics, vol. 24, no. 4, pp. 471–479, 2019.
[40] H. Sarojadevi, J. Meghashree, and D. Shruthi, "IoT Based System for Alerting Forest Fire and Control of Smuggling," International Journal of Advance Research and Innovation, vol. 7, no. 2, pp. 91–93, 2019.
[41] J. Scholz, A. De Meyer, A. S. Marques, T. M. Pinho, J. Boaventura-Cunha, J. Van Orshoven, et al., "Digital Technologies for Forest Supply Chain Optimization: Existing Solutions and Future Trends," Environmental Management, vol. 62, no. 6, pp. 1108–1133, 2018.
[42] V. Nicheporchuk, I. Gryazin, and M. N. Favorskaya, "Framework for Intelligent Wildlife Monitoring," In International Conference on Intelligent Decision Technologies, pp. 167–177, Springer, Singapore, June 2020.
[43] G. S. Karthick, M. Sridhar, and P. B. Pankajavalli, "Internet of Things in Animal Healthcare (IoTAH): Review of Recent Advancements in Architecture, Sensing Technologies and Real-Time Monitoring," SN Computer Science, vol. 1, no. 5, pp. 1–16, 2020.
[44] M. G. Karthik and M. B. Krishnan, "Hybrid Random Forest and Synthetic Minority Over Sampling Technique for Detecting Internet of Things Attacks," Journal of Ambient Intelligence and Humanized Computing, 1–11, 2021.
[45] E. D. Ayele, N. Meratnia, and P. J. Havinga, "Towards a New Opportunistic IoT Network Architecture for Wildlife Monitoring System," In 9th IFIP International Conference on New Technologies, Mobility and Security (NTMS), pp. 1–5, IEEE, February 2018.
[46] Anonymous, Instant Detect 2.0 emerges. Accessed from Instant Detect 2.0 emerges | WILDLABS.NET. (n.d.). https://www.wildlabs.net/resources/case-studies/instant-detect-20-emerges.
[47] Anonymous, Trail guard-RESOLVE. Accessed from https://www.resolve.ngo/trailguard.htm.
[48] P. Cipresso, I. A. C. Giglioli, M. A. Raya, and G. Riva, "The Past, Present, and Future of Virtual and Augmented Reality Research: A Network and Cluster Analysis of the Literature," Frontiers in Psychology, vol. 2086, 2018.

[49] W. S. Kim, W. S. Lee, and Y. J. Kim, "A Review of the Applications of the Internet of Things (IoT) for Agricultural Automation," Journal of Biosystems Engineering, vol. 45, no. 4, pp. 385–400, 2020.
[50] P. Liu, Y. Zhang, H. Wu, and T. Fu, "Optimization of Edge-PLC-Based Fault Diagnosis With Random Forest in Industrial Internet of Things," IEEE Internet of Things Journal, vol. 7, no. 10, pp. 9664–9674, 2020.

Index

https://doi.org/10.1515/9783110785234-012

Also of Interest

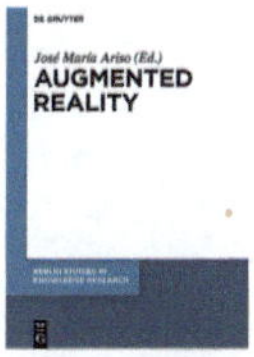

Augmented Reality
Reflections on Its Contribution to Knowledge Formation
José María Ariso (Ed.), 2017
ISBN 978-3-11-049700-7, e-ISBN 978-3-11-049765-6

Artificial Intelligence for Virtual Reality
Volume 14 in the series De Gruyter Frontiers in Computational Intelligence
Anett Jude Hemanth, Madhulika Bhatia and Isabel De La Torre Diez (Eds.), 2023
ISBN 978-3-11-071374-9, e-ISBN 978-3-11-071381-7

Interacting with Presence
HCI and the Sense of Presence in Computer-mediated Environments
Giuseppe Riva, John Waterworth and Dianne Murray (Eds.), 2014
ISBN 978-3-11-040967-3, e-ISBN 978-3-11-040969-7

Advances in Industry 4.0
Concepts and Applications
M. Niranjanamurthy, Sheng-Lung Peng, E. Naresh, S. R. Jayasimha and Valentina Emilia Balas (Eds.), 2022
ISBN 978-3-11-072536-0, e-ISBN 978-3-11-072549-0

Soft Computing in Smart Manufacturing
Solutions toward Industry 5.0
Tatjana Sibalija and J. Paulo Davim (Eds.), 2022
ISBN 978-3-11-069317-1, e-ISBN 978-3-11-069322-5